OIL OR WATER . . .
ONLY ONE IS GOOD TO THE
LAST DROP

OIL or WATER

ONLY ONE IS GOOD TO THE LAST DROP

A WORK OF SCIENCE FACTION

BY JOHN E. HORNER

To order additional copies of this book, contact:
Xlibris Corporation
1-888-795-4274
www.Xlibris.com
Orders@Xlibris.com
110740

Table of Contents

Chapter 7

Chapter 8

Chapter 12

Acknowledgements

I would like to acknowledge the following people who have contributed time, energy, patience, and constructive criticism in the making of this book.

My Children:
Wendy and Husband Mike, John, Aaron and fiancée Lisa, Caleb, Amber and Husband Mike

Two Sisters:
Ann, who published grandfather's diary and Grace, who was with me when the book idea became resolve.

Two brother in-laws:
Paul and Ron who supplied ample, factual data, on multiple sides of the question.

Especially Kristina Marriott who has been sparkplug, and accelerator, as well as my computer guru in assembling this book.

My friends:

Dick Brighthaupt, a tireless internet researcher who supplied much of the factual data And Bob Kennedy who was a great listener as I processed my thoughts in the creation of this book.

A belated acknowledgement to Yale Divinity School: who encouraged me to speak my mind on ethical issues, although not fully manifested until after my 55th birthday.

My Sunday School Class, for enduring my obsession with relevant facts.

My Prayer Group, for their mature helpful patience and insights.

Jeremy Peña, for his creative contribution by way of an excellent representation of my vision shown in the artwork he produced as a cover for this book

Many additional scholarly works are excerpted, acknowledged, and praised in the body of the manuscript.

Large picture of the west coast of the United States, courtesy of NASA.

Preface

While sitting talking to John about his book on global warming he asked me to tell him what I have learned from our many conversations.

I have learned this . . . Through the replacement of water in various shapes and size of containment and in various locations in the Sahara desert, it is possible to change the precipitation factor enough to create a green paradise in a now very dry and waterless climate. This in turn, could very possibly change the precipitation factors in our climate altogether. If we multiply our containment placement to other areas in the world, we could possibly change the global warming conditions forever.

I have learned that the oil spills that occurred in the 80's, greatly affected the plankton in our oceans and that the plankton levels greatly affect the precipitation that naturally occurs under the right conditions. It is this lack thereof that has slowed down the replacement of water into our climate, causing a great deal of the global warming we are experiencing today. I have learned that it has been man's neglectful and wasteful ways that have caused much of the issues and if we would just choose to join forces, we may very well be able to reverse the damage we have done.

Though many may think these are very out there and crazy ideas that John has come up with, I say . . . I am sure many thought Benjamin Franklin was nuts, when he flew his kite in the middle of a Lightning Storm.

Genius is as genius does!
Kristina A. Marriott

Foreword

I was deeply engaged in my writing before I sensed what I was writing about. My original intention was to explore fully the intense variety of essays and studies in the fecund area of Global Warming-Climate Change. I can admit that it now proliferates at a rate faster than I can read. But I found something missing from the discussion. The missing element is **Color**—not the Black, White, Red, Bronze or Yellow that fills the human interest sections of newspapers. This is the Black and Green of Nature-earth and ocean. Even the murky green of large segments of the ocean is arrayed against the deep blue transparency of other parts of the ocean. The opposite on land and in the air is the refractive silver sand of the Sahara, the golden hue of wheat harvest and the flashy coats of autumn. Coupled with the intense refractive nature of ice and snow and the indescribable silvers and blacks of thunderstorms, even the ocean surface (aside from vegetative murkiness) has amazing contrasts. This is taken from 2008 World Almanac pg. 295 pertaining to ultraviolet {UV} forecast—Reflectivity. Reflective surfaces intensify UV exposure. Grass reflects 2.5%-3% of UV radiation reaching the surface: sand 20%-30%; snow and ice 80%-90%; and water up to 100% {depending on reflection angle}. The Almanac also states clear skies allow 100% UV transmission to surface, broken clouds allow about 73% and overcast conditions allow 32%. I can't help myself from musings that some {%?} UV must be good for humans and plants?

These musings should have special relevance when we consider that the older charts of atmospheric content show water vapor as a variable constituent of the atmosphere estimate—1%-4%. The newer charts (after 1978-1983 releases of equatorial oil) only allow 1%-2% atmospheric water content. Some even stoop to 0-2 %

All of the foregoing colorful material forms a formidable task for the narrow confines of any assigned climatologic study. Even jet contrails and soot from diesel trucks and trains exhaust play a significant role in the albedo balance of the earth. I will attempt to evaluate the Ozone Hole(s) for its long term consequences of the reflective-absorptive imbalance of the earth. This imbalance is made up of problems and opportunities. The problems result from the universal dun grey that provides insufficient motivation for the climate engine. The opportunities are born from sea and land, black and white, {snow and mud} reflective and absorptive. We dare not forget the role of dust—especially for its many tiered reflective and absorptive capacities. The physical force of rain-soaked black earth when exposed to the full rays of the summer sun is something that no farmer ever forgets. Nor can he forget, the absorptive cool green of a chest high field of growing corn.

Am I giving away the secrets of the plot? But for the climate modeler these colors might qualify as **The Elephant in the Kitchen,** but there are other candidates since some forecasters already perceive the elements of color in some of their forecasts. If this work should accompany an Abraham-like exodus to the higher more difficult fields, it just might work if the respect for water (retention and absorption) is held as a **holy calling.** The theme of that holy calling is to be labeled Re-raining. For proof I cite this amazing statistic; only 50% of Amazon rain comes from the ocean! The other 50% comes not from the Caribbean or the Pacific but from transpiration from trees and grasses. **My ethical calling is to fix 'what's broke', and it might just extend to caring about what is about** to break.

I have found some material to provide historical perspective to focus us on our collective goals. That collective assurance is mostly negative;

but saying it aloud allows me to announce my own purposes in harmony (hopefully) with the long term collective objective. Here goes!

From *Guns, Germs, and Steel the fates of human societies* **pp. 410-411** (and other selections). Jared Diamond: "Why, then, did the Fertile Crescent and China eventually lose their enormous leads of thousands of years . . . (ultimate cause): For the Fertile Crescent the answer is clear. Once it had lost the head start that it had enjoyed thanks to its locally available concentration of domesticable wild plants and animals, the Fertile Crescent, possessed no further compelling geographic {sic}. The disappearance of that head start can be traced in detail, as the westward shift in powerful empires. After the rise of Fertile Crescent states in the fourth millennium B.C., the center of power initially remained in the Fertile Crescent, rotating between empires such as those of Babylon, the Hittites, Assyria, and Persia. With the Greek conquest of all advanced societies from Greece east to India under Alexander the Great in the late fourth century B.C., power finally made its first shift irrevocably westward.

"The major factor behind these shifts became obvious as soon as one compares the modern Fertile Crescent with ancient descriptions of it. Today, the expressions **Fertile Crescent** and **World leader in food production** are absurd. Large areas of the former Fertile Crescent are now desert, semidesert, steppe, or heavily eroded or salinized terrain unsuited for agriculture. Today's ephemeral wealth of some of the regions nations, based on the single (non-renewable)[*That is his choice of words, throughout this book I will show how fundamentally wrong this word non-renewable is and will attempt to highlight the handiest most easily accessible source of renewing.*] resource of oil, conceals the regions long-standing fundamental poverty and difficulty in feeding itself.

'In ancient times, however, much of the Fertile Crescent and eastern Mediterranean region, including Greece, was covered with forest. The region's transformation from fertile woodland to eroded scrub or desert has been elucidated by paleobotanists and archaeologists.

17

Its woodlands were cleared for agriculture, or cut to obtain construction timber, or burned as firewood or for manufacturing plaster. Because of low rainfall and hence low primary productivity (proportional to rainfall), regrowth of vegetation could not keep pace with its destruction, especially in the presence of overgrazing by abundant goats. With the tree and grass cover removed, erosion proceeded and valleys silted up, while irrigation agriculture in the low-rainfall environment led to salt accumulation. These processes, which began in the Neolithic era, continued into modern times. For instance the last forests near the ancient Nabataean capital of Petra, in modern Jordan, were felled by the Ottoman Turks during construction of the Hejaz railroad just before World War I.'"

A lengthy Search for an acceptable norm to return to—begins with a look backward here.

Lest the readers confusion remains centered upon the Fertile Crescent and the slow systemic failures that began there, I intend to take you back one step further to the one-time viability of the Sahara Desert in order that a deeply rooted repentance can give a proper and enduring foundation to the fundamental reforms that are needed. Same source pp. 386, ff.

"When Europeans reached sub-Saharan Africa in the 1400's, Africans were growing five sets of crops, each of them laden with significance for African history. The first set was grown only in North Africa, extending to the highlands of Ethiopia. North Africa enjoys a Mediterranean climate, characterized by rainfall concentrated in the winter months. (Southern California also experiences a Mediterranean climate, explaining why my basement and that of millions of other southern Californians often gets flooded in the winter but infallibly dries out in the summer.) The Fertile

Crescent, where agriculture arose, enjoys that same Mediterranean pattern of winter rains.

"Hence North Africa's original crops all prove to be ones adapted to germinating and growing with winter rains, and known from archaeological evidence to have been domesticated in the Fertile Crescent beginning around 10,000 years ago. Those Fertile Crescent crops spread into climatically similar adjacent areas of North Africa and laid the foundation for the **rise of Egyptian civilization.** They include such familiar crops as wheat, barley, peas, beans, and grapes. These are familiar to us precisely because they also spread into climatically similar adjacent areas of Europe, thence to America and Australia, and became some of the staple crops of temperate-zone agriculture around the world.

"As one travels in Africa across the Saharan desert and reencounters rain in the Sahel zone just south of the desert, one notices that Sahel rains fall in the summer—**{Author's note, these rains might properly have their origination in the powerful direct and increasingly penetrating sunlight that warms the South Polar oceans in what is increasingly described in ominous terms by Antarctic based climatologists, thus to me it is no coincidence that the decline of the ozone layer, and its corollary—the sudden and unseemly heating of the South Atlantic ocean—occurred in tandem with the decimation of the Sahel}— rather than in the winter (your understandable confusion can remain until dealt with later.)** "Even if the Fertile Crescent crops adapted to winter rain could somehow have crossed the Sahara, they would have been difficult to grow in the summer-rain Sahel zone. Instead, we find two sets of African crops whose wild ancestors occur just south of the Sahara, and which are adapted to summer rains and less seasonal variation in day length. One set consists of plants whose ancestors are widely distributed from west to east across the Sahel zone and were probably domesticated there. They include, notably, sorghum and pearl millet, which became the staple cereals of much of sub-Saharan Africa. Sorghum proved so valuable that it is now grown in areas with hot, dry climates on all the continents, including the United States."

{Authors note, because Americans under 40 are probably preconditioned to think of the Sahel in terms of AIDS, starving cattle, genocide etc., I will later submit rational rainfall data along with passing reference to Dr. Albert Schweitzer's lengthy success at Lambarene, south of the Sahel. Included will be the war-time famine induced by excessive rainfall.}

"The other set consists of plants whose wild ancestors occur in Ethiopia and were probably domesticated there in the highlands. Most are still grown mainly just in Ethiopia and remain unknown to Americans-including Ethiopia's narcotic chat, its banana-like ensete, its oily noog, its finger millet used to brew its national beer, and its tiny-seeded cereal called teff, used to make its national bread. But every reader addicted to coffee can thank ancient Ethiopian farmers for domesticating the coffee plant. It remained confined to Ethiopia until it caught on in Arabia and then around the world, to sustain the economies of countries as far-flung as Brazil and Papua, New Guinea.

"The next-to-last set of African crops arose from wild ancestors in the wet climate of West Africa. Some, including African rice, have remained virtually confined there; others, such as African yams, spread throughout other areas of sub-Saharan Africa; and two, the oil palm and kola nut, reached other continents. West Africans were chewing caffeine-containing nuts of the latter as a narcotic, long before the Coca-Cola Company enticed first Americans and then the world to drink a beverage originally laced with its extracts.

"The last batch of African crops is also adapted to wet climates but provides the biggest surprise. Bananas, Asian yams, and taro were already widespread in sub-Saharan Africa in the 1400's and Asian rice was established on the coast of East Africa. But those crops originated in tropical Southeast Asia. Their presence in Africa would astonish us if the presence of Indonesian people on Madagascar had not already alerted us to Africa's prehistoric Asian connection. Did Austronesians sailing from Borneo land on the East African coast, bestow their crops on grateful African farmers, pick up African fishermen, and sail off into the sunrise to colonize Madagascar, leaving no other traces in Africa? . . ." (from Guns, Germs, and Steel . . . pp. 386,ff)

North Africa, can there be a connection to a million year Ice Age?

You, dear reader have every right to ask, "Why besiege me with ancient bits of irrelevant data?" The answer is that the material when properly understood, illuminates the reason that we have been trapped in a (million?) years of recurring ice age (s)? In its baldest terms the location of Africa with the Sahara straddling the equator insures that the earth under tropical desert conditions can divert a crucial amount of the sun's radiance back into space. I don't expect anyone to grasp this concept because I didn't when I first read it in an American Academy of Science report. A few years later when I read of the progressive widening of the Sahara on its southern edge, it still didn't penetrate. But last year when I visited a (Commune?) in southern Missouri, the senior man claimed that with [60?] inches of annual rainfall, they couldn't make a crop (without irrigating) because the gap between spring and fall rainfall had become too great. By my old Nebraska standards, 16 inches of timely rain could sometimes produce 100 bushels of corn. Then and only then did it dawn on me that the increasing dryness of the Sahel might be related, by way of reducing the global wind circulation patterns to the absence of enough air movement to drive Atlantic and Gulf moisture ashore. We are interested in regular and timely rainfall on the continents, so, we begin this book with a safe look at the distant past, and there we do not want just any rainfall, but we want rainfall where man's activities have interrupted past rainfall. In other words, we are attempting not just any rainfall, but to renew the rainfall that has been interrupted by man's activities in the past. A consensus view of that time will give a perspective that allows us to chart a course carefully amid the wildly varying opinions of today.

Chapter 1

GLOBAL WARMING IN PURSUIT OF A $25 MILLION DOLLAR PRIZE COURTESY OF SIR RICHARD BRANSON

A personal introduction

I first blundered headlong into global warming in the summer of 1980. I had plenty of warning. Two years earlier I had purchased, read and reread a book titled, *The Weather Conspiracy—The coming of the New Ice Age.* Pub. 1977. This book cited abundant evidence (purportedly) with the help of the CIA and data from Greenland ice cores covering climate evidence going back 150,000 years. In the back of my mind arose the conviction, (actually it's mentioned in the book) that ice ages recur with regularity every 100,000 years and perhaps last as long as 90,000 years. How much of this was hype or fear mongering I haven't decided.

From Weather Conspiracy p. 201. "There is, moreover, growing consensus among leading climatologists that the world is undergoing a cooling trend. If it continues, as feared, it could restrict production in both the USSR and China among other states and could have enormous impact, not only on the food-population balance, but also on the world balance of power.

This paper reviews present trends and projections for world population and food production under assumptions of normal weather, and then assays a necessarily tentative exploration of the ramifications of a cooling climate. A final section addresses the political and other implications for the U.S. of its potential role as the main food exporter in an increasingly hungry world."

If I needed any further convincing, one morning in 1978, I stepped onto the Yellow Freight dock, (open to the elements), and the chill factor was broadcast as 78 degrees below zero. I knew it was bad because no work was being done—men were gathered—huddled near any heat source. If memory serves me correctly, we had snow cover from mid-December until February. 1979 was not much better—consequently by 1980 I had burned up most of my baby fat. I noted in passing that there was a runaway oil well in the Gulf of Mexico called Ixtoc I.

Because I had breathed the academic air, I labored under the assumption that we had had our good 10,000 years. Oh Boy! Was I surprised by the summer of 1980? Spring came early and more than pleasantly warm. Summer was hot, hot, hot, **hot!** After 27 years I can still remember with vivid clarity the day that I was assigned a PROLONGED STINT IN A TARPAULIN COVERED TRAILER—black tarp to you solar conversion enthusiasts. The temperature on the open dock was 105° degrees that day—one of a succession of six weeks that the daytime highs exceeded 100°. When I returned to the dock I noted with considerable alarm that I had goose bumps on both arms as well as chills running up my back. What, dear reader, do you think that temperature was inside that trailer?

A Significant Experiment Related To
Oil and Evaporation

That summer was an all-time high for K.C., Missouri, (comparing only to 1936) There were 115 heat related deaths, locally, and 1265 deaths nationally, figs. from 2008 World Almanac. And because I had a curious disposition—I did a little experiment. I took 11 plastic milk jugs and cut the tops off of them. I put a precise amount of water into each jug. One was plain for a control. Each of the others had a different substance in it. Of unusual significance were these four containers the 1st with a teaspoon of new oil, the 2nd a tablespoon of new motor oil, the 3rd a teaspoon of used motor oil, and the 4th a tablespoon of used motor oil. In the others I placed an assortment of items that could alter the evaporation rate. After a few days I checked it. Everything in the jugs proceeded to evaporate at the common sense expected rate with the exception of the jugs containing used motor oil. After allowing a few more days of evaporation, I concluded the used motor oil forms an almost impenetrable film on top of the water. Unexpectedly, the new oil clumped and showed less than expected evaporation impedance.

If there is a conclusion to be reached, one possibility is that the miniscule amount of motor wear contributed a magnetic repulsion to the film. Of course I had no capacity at that location and time to evaluate the influence of acids. It is only a flimsy theory but the reality of the experiment should have some relevance to the common sense world that I intend to dwell on for at least the next few pages. This theory has less than a full body of knowledge behind it because I was unable to compare the evaporation rate of water from beneath a film of fresh-drilled crude. Even if I had some fresh crude it would provide a less than adequate basis for generalization. Not only are there multiple crudes but they have been spilled in multiple conditions of the ocean. See enclosed fact sheets on the next 2 pp.

		Major Oil Spills
2002 November	Spain	Prestige carrying 20 million gallons (70,000 metric tons) of fuel oil broke up off the Spanish coast.
2001 January	Ecuador	Ecuadorean-registered ship Jessica, spilled 175,000 gallons of diesel and bunker oil into the sea off the Galapagos Islands in what was seen as one of Galapagos' worst environmental disasters.
2000 June	South Africa	Some 1,400 tonnes of heavy fuel oil leaked from the bulk carrier Treasure off Cape Town prompting massive rescue of Jackass penguins on Dassen and Robben Islands.
2000 January	Brazil	A ruptured pipeline spewed about 340,000 gallons of heavy oil into Guanabara Bay, Rio de Janeiro, in Brazil.
1999 December	France	The stern of the Maltese tanker Erika sank off the northwest of France after splitting in two. It was carrying 25,000 tonnes of viscous fuel oil.
1997 December	Sea of Japan	Japan Russian tanker Nakhodka spilled 19,000 tonnes of oil after breaking in two in the Sea of Japan.
1996 February	UK	UK Liberian-registered Sea Empress hit rocks near Milford Haven, Wales, spilling 72,000 tonnes of oil.
1994 October	Portugal	Panamanian tanker, Cercal, spilled about 2,000 tonnes of crude into the sea after striking a rock near Leixoes harbour, in Oporto
1994 March	United Arab Emirates	15,900 tonnes of crude oil leaked into the Arabian Sea after the Panamanian-flagged Seki collided with the UAE tanker Baynunah 10 miles off the UAE port of Fujairah.
1994 March	Thailand	About 105,000 gallons of diesel fuel spilled into the sea four miles off the eastern Sriracha coast after oil tanker Visahakit 5 and a cargo ship collided.
1993 January	UK	The tanker Braer hit rocks near the coast of the Shetland Islands shedding its cargo of 85,000 tonnes of crude in the worst oil wreck in British waters for 26 years.
1992 December	Spain	Greek tanker Aegean Sea ran aground and broke in two near La Coruna spilling most of its 80,000 tonne cargo of oil.
1992 September	Indonesia	Liberian-registered tanker Nagasaki Spirit collided with container Ocean Blessing in the Malacca Straits spilling some 12,000 tonnes of crude.
1991 May	Angola/ Liberia	A Liberian-registered supertanker, ABT Summer, leaked 260,000 tonnes of oil after an explosion off Angola causing an oil slick 17 nautical miles by three.
1991 April	Italy	The Haven spilled more than 50,000 tons of oil off Genoa in Italy.
1991 January	Persian Gulf	Iraq released about 460 million gallons of crude oil into the Persian Gulf during the Gulf War.
1990 February	USA California	The tanker, American Trader, leaked 300,000 gallons of crude from a gash in the hull causing an oil slick 14 miles long polluting Bosa Chica, one of southern California's biggest nature preserves
1989 December	Morocco	After explosions and a fire Iranian tanker Kharg-5 was abandoned spilling 70,000 tonnes of crude oil, endangering the coast and oyster beds at Oualidia.
1989 March	USA Alaska	Exxon Valdez grounded and spilled 10 million gallons (38,800 tons) of crude oil into Prince William Sound in Alaska.
1983 August	South Africa	Fire broke out on the Spanish tanker Castillo de Bellver and 175.6m gallons of light crude burnt off the coast at Cape Town. Fire broke out on the the Castillo de Bellver and its cargo of 252,000 tonnes of oil burnt.
1979 July	Trinidad	160,000 tons of crude oil spilled after a collision off Tobago between the Atlantic Empress and the Aegean Captain.
1978 March	France	Amoco Cadiz encountered stormy weather and ran aground off the coast of Brittany, France on March 16, 1978. Its entire cargo of 68.7 million gallons (220,000 tons) of oil spilled into the sea, polluting about 200 miles of Brittany's coastline.
1977 February	Northern Pacific	Liberian-registered Hawaiian Patriot caught fire in the Northern Pacific spilling 30.4m gallons.
1976 December	USA Massachusetts	December 15, 1976, the Argo Merchant ran aground on Fishing Rip (Nantucket Shoals), 29 nautical miles southeast of Nantucket Island, Massachusetts in high winds and ten foot seas. Six days later, the vessel broke apart and spilled its entire cargo of 7.7 million gallons of No. 6 fuel oil, causing a slick 100 miles long and 60 miles wide.

Compare & contrast 713,000 tons spilled from 1989 to 2002 excluding Persian Gulf 1991. With the record from the 2001 World

Almanac pg. 259, 5 spills & releases from March 16, 1978 to February 1983 totaled 1,973,000 tons of oil spilled mostly on equatorial oceans!

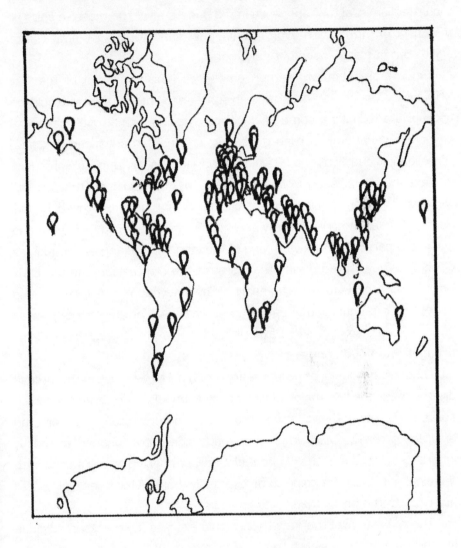

Even though my attempts to understand the consequences of massive spills on the world's oceans seem pitifully crude to me—they are better than the dismissal that appeared in the papers after the massive spills at the onset of the 90 Gulf War. In brief, the "sandstorm sunk em." Out of sight—out of mind—does not quite get it.

Since I am about to launch on the recharacterization of the last 85 years of weather and climate history—esp. man's role, it would be appropriate to outline some of my embarrassing lack of qualifications—particularly academic, I read a lot, but, possess no advanced degrees. Of colloquial qualifications, 35 years working on an open freight dock in Kansas City, Missouri, was preceded by 12 years on the farm in Nebraska 1945-57. Of interest—the families that were linked by the Sept. 2nd 1929 marriage of my parents represented both Oklahoma and Nebraska persons of the cloth and of the land. The distinguishing feature of both is they stayed during the worst of the Depression and the Dust Bowl. Their stories furnish the heart of my motivation to restore the land as well as the courageous backdrop of what I will attempt to say. I was born in 1938, so by 1945 I experienced the bright side of Great Plains **drought and flood** contrast.

I chose a colloquial or personal approach partly because of the intense divergence of the two major viewpoints. Additionally there is no concerted agreement upon either the temperature or the projected future or even of present reality. Most important, both sides have chosen to mostly ignore the elephant in the kitchen. Not only am I going to call attention to the elephant, but I am going to be expressly thankful for it, as well as call attention to the problems.

There are four authorities that I would express my thanks in advance for their particular insights to global warming. Al Gore for elevating the subject into the public mind. Chris Horner for his fact-filled exposition of possible private agendas titled The **Politically Incorrect Guide to Global Warming.** Robert W. Felix took me on a sojourn through time including prehistory in his book titled, **Not by Fire but by Ice.** His work placed the subject beyond the vitriol and hysteria of less historically grounded individuals. And last but not least George Bush Sr. for presiding over the CIA during the time that they contributed to the making of the book. *The Weather Conspiracy; The Coming of the New Ice Age.*

A fifth author Ritchie Calder wrote a book titled; *After the Seventh Day —the world that man created.* His work may have placed into my intellect the conviction that I already had in my gut— that man can make a difference in his environment. That it will be a positive difference is the reason for some extra lengths that I take to expound on some fairly simple ideas. But none of these authorities are responsible for the ideas in the following paragraphs.

The reader will note that I have cited two books predicting an ice age to balance the rather large political weight of Al Gore. Because I am an activist by nature and an optimist by bent I intend to spend some extra effort seeking balance. Sir Richard Branson sounds like a kindred spirit. It is my intention to provide him with a rationale that will permit him to act with great confidence as well as great caution.

Now to the tale of The Elephant in the Kitchen.

In 1922 or 23 a runaway oil well in lake Maracaibo, Venezuela spilled according to one source a hundred thousand barrels of oil a day for nine days. My source was not as definitive as the Encyclopedia Britannica which lists yearly production in the region at 10,000,000 bbls. It makes no mention of the spill but this is a 1965 edition before spills were recognized as endemic to the process.

Another source shows the beginning of Hurricanes (and the Gulf Stream?) to be the East of the Atlantic near Africa. From there they accompany the Gulf Stream as it proceeds Westward & along the North Coast of South America and then Northeast to its ultimate surface destination Norway. That circuit encompassed & spread most of that Maracaibo 900,000 bbl. spill. *That early form of ocean sealing resulted in the epochal 1927 Mississippi flood.* Frank Schatzing in The Swarm delineates a 1,000 year circuit of the water currents around the globe attempting in a few poetic paragraphs to capture the swirls and eddies, the salt laden sinking as well as the polar uprising. But for our purposes only the reference to the eastward movement out of the Gulf of Mexico is significant. There it becomes {clearly labeled} the Gulf Stream and picks up enormous quantities of heat before heading for Labrador and Norway. But I am a bit of a Budykoyan & Thermal energy determines much of the course of water. We'll follow that idea through the murk of living things as well as the mess of oil residue. The Budykoyan theory will be amplified a bit later in the text.

On the plane to Norway I picked up a tourist magazine claim that the Gulf Stream provides more energy than 3,000 nuclear power generating stations. I didn't assign the claim much credibility until I experienced temperate zone trees and fields of grain north of the Arctic Circle.

At any rate 900,000 barrels of oil must have taken a while to depart the shallow sandbar that guarded Lake Maracaibo's merger with the Gulf Stream. This is 1922 well before any scientific efforts to clean up.

My relative from Oklahoma assures me that streams in Oklahoma ran oil. I suppose that means that there were periods when the water was hardly visible. But this is by no means a legal brief nor the prelude to a vindictive lawsuit. Since they were of my approximate age, I suppose that means that even in the forties there were not significant efforts to prevent messes.

Another source of authority suggests that the three main areas of the 1920s world of oil were Venezuela, Baja California & the three state areas Texas, Oklahoma and Louisiana. Almost no one that I know of suggested that the inundation of the Gulf of Mexico with a blanket of oil produced the Dust Bowl. *And-or the 1927 super-flood which in all probability shredded a lot of root systems which then allowed the sun to work its hardening magic on the naked exposed soil!* Plenty of blame was assigned by the print media & governmental authority particularly conservationists speaking to farmers and farming techniques.

Chapter 2

By now it should be apparent to the over-eager conclusion-jumper. This is just another diatribe against Big Oil. Far from it! By the time I have completed this manuscript you will see abundant reasons to allow the phrase Global Warming, to fall into disuse. *{Author's interjection: this material is old, but in writing today(early 2010) The phrase Global Warming has been dropped from the newspapers more completely than yesterday's sex scandals, as people shiver quietly next to their thermostats.}* It will be replaced by the narrowly descriptive phrase, Ocean Sealing. We might even learn to be cautiously grateful for two of their potentially key contributions to forestall Climate Tipping. If the phrase Global Warming is to be retained it should be cautiously applied (in a strictly temporary way) to that portion of the electromagnetic spectrum that penetrates the ozone (Hole) (s). For I truly believe that this hole is temporary (as well as periodically recurring on a bi-decadal basis naturally,) and can be circumscribed whenever we are ready to cooperate with those consequences (opportunities) that result.

Ozone Hole and Chlorine and Both Poles (first discussion)

There are questions to be asked about this area of the world, primarily Antarctica and about chlorine. I'm allowing my bias to show a bit here. Why should CFCs be the primary culprit in the distinctly American search for villains? I personally grew up free of infectious diseases while drinking untreated Nebraska well water. In Kansas City I acquired a mysterious water allergy that was not fully overcome until I began to prepare a huge glass of water for my bedside—(the only water of the day) to be consumed during and after the wee hours of the morning. Such was my confusion that I believed that I was over-reacting to the temperature at the tap. Finally someone explained that tap water meant death to goldfish and had to be set out for a period of time to allow the toxic chlorine to dissipate. On the other hand, as a youth I rejoiced in the smell of ozone after a thunderstorm. It never occurred to me that ozone might be toxic—as long as the lightning bolt missed me. I further believe there are societies (not just water-bottlers) which use ozone exclusively for water purification and their health and longevity compare favorably with ours. My opinions don't carry enough weight, even to persuade me, so, here are some facts from the internet.

Ozone Applications. Industrial applications At present, the uses of ozone as an industrial chemical are somewhat limited. The largest use of ozone is in the proper preparation of pharmaceuticals, synthetic lubricants, as well as many other commercially useful organic compounds, where it is used to sever carbon carbon bonds. It can also be used for bleaching substances and for killing microorganisms in air and water sources. Many municipal drinking water systems kill bacteria with ozone instead of the more common chlorine. Ozone has a very high oxidation potential. Ozone does not form organochlorine compounds, nor does it remain in the water after treatment so some systems introduce a small amount of chlorine to prevent bacterial growth in the pipes, or may use chlorine intermittently, based on results of periodic testing. Where electrical power is abundant, ozone is a cost-effective method of treating water, as it is produced on demand and does not require transportation and storage of hazardous chemicals. Once it has decayed, it leaves no taste or odor in drinking water. Low levels of ozone have been advertised to be of some disinfectant use in residential homes, however, the concentration of the ozone required to have a substantial effect on airborne pathogens greatly exceed safe levels recommended by the US occupational safety and health administration and environmental protection agency. {Among the many uses of ozone}

Ozone is used in spas and hot tubs to kill bacteria in the water and to reduce the amount of chlorine or bromine required by reactivating them to their free state. As ozone does not remain in the water long enough, also ozone by itself is ineffective at preventing cross contamination among bathers and must be used in conjunction with these allergens. Gaseous ozone created by ultraviolet light or by corona discharge is injected into the water. Ozone is also widely used in treatment of water and aquariums and fish ponds. Its use can minimize bacterial growth, control parasites, and reduce or eliminate yellowing of the water. Because it decomposes rapidly, the ozone has no effect on the fish at properly controlled levels.

It is not illegal to sell medical grade ozone machines in the US, nor is it illegal to own one or use one. What is illegal is to sell them while claiming it treats disease. Many people use ozone therapy in the US despite its unrecognized status with the FDA and the medical establishment. It is legal to sell or own a medical grade ozone machine in the US. It is also legal to administer ozone to oneself." Retrieved from "http://en.wikipedia.org/wiki/Ozone

After these facts it remains only to deal with public perception of the ozone crisis. In other words what is hidden in plain sight. Some of that is covered by the intro to this book and reference to the world almanac and book of facts. The wonderful work of Susan Solomon will be considered later.

At any rate, to be genuinely cautious we could purify our water with ozone and hope that any overuse escaped upwards instead of the present system where I feel slightly guilty whenever I use CLOROX.

If we were strictly rational about the Ozone Hole, we might have a statistical presentation of the number of days that its influence permeated, as well as the number of square miles over which it held sway! By the way, I've found some of these statistics & can hardly wait to indulge in Limited praise of Susan Solomon (later). The other side of the same coin could show the number of days that snow cover predominated as well as the number of square miles that were covered by snow and ice. The universal nature of the relationship should help to clarify some of the mysteries of Global Warming especially the contrast of stratospheric absorption vs. incursion with sea level reflection (snow ice-clouds-) vs. phytoplanktonic absorption and subsequent evapo-transpiration.

But I won't be personally satisfied with any study that does not encompass two Solar cycles (44 years) with at least a minimal perception of natural (solar cycle caused) destruction of ozone. The understanding of solar cycles under girding climate theories at the time of the Dust Bowl might help us to distinguish between a natural vs. man-caused ebb and flow.

But enough light-weight opinions—let's have some facts: from *http://science.nasa.gov/headlines/y2001/astlloct_1.htm.* 'In years when planetary waves (or long waves) in the Northern Hemisphere are unusually weak, an ozone hole can form over the Arctic! The Himalayan plateau is a terrific forcing function for these waves in the north,' says Paul Newman an atmospheric physicist at NASA's Goddard Space Flight Center. 'If you didn't have the Himalayas, *the stratosphere over the Arctic would be much colder than it is.'*

[*Authors emphasis added*]

'The graph accompanying these words shows a 1984 north pole with strong long waves and weaker than normal stratosphere and less ozone loss. 13 years later 1997 shows weak long waves and colder than normal stratosphere and more ozone loss.' (And now to further pursue the elephant in the kitchen). By the way, Sir Richard, I must apologize for putting you thru this pre-protocol. My conscience prohibits licensing you to make these simple but far-reaching changes without knowing the caveats or potential side effects of this noble effort. I have met you only briefly via TV but I am prepared to like you and your effort. Like the "Six Blind Men of Hindustan" there are multiple descriptions of the Elephant.

Personal Farm Introduction

Before I launch into something that will test the uninitiated, let me share with you from the ordinariness of life on the farm in Nebraska circa 1945 to 1949. However, it was anything but ordinary. My father was a man of many jobs in the '30's none of them paid much. (Due to parishioner lack of money to pay him,) He was dismissed from a preaching job in 1933 on a Sunday that the noontime temperature registered 133 degrees. From that transition he planted trees alongside Hwy. 6. He was partner in a John Deere dealership in 1938 the year I was born. There was no money in it so he graduated to selling Fuller Brushes during the War.

During this period of underemployment his mind and spirit were not idle. Many magazines and periodicals dwelt on the shortcomings of Great Plains farming. One of the most striking and persistent themes was the 22 year solar cycle. In brief it described a natural boom and bust cycle inevitably bringing drought every 20 or 22 years. One of the staples of the theory was an increasing invasion of ultraviolet radiation. This signaled the nadir of rainfall because the ultraviolet radiation put a halt to the microscopic algal growth that beclouded the ocean. Under normal conditions, the proliferation of algae created a natural El Nino (Surface warming) effect that encouraged the evaporation so necessary for rainfall to occur on land.

Spring of 1945 arrived with abundance of moisture. The horses pulling a hay rack load of furniture were driven up the ditch because the mud road was impassable. We arrived onto the farm which didn't support a family of four, with high hopes that it would bring prosperity to a family of eight children eventually to be ten in number. My grandmother on moving to town spoke the cautionary words," *You are moving from low ground to high ground, from the moister East to the higher and dryer West, Don't get your hopes too high.*"

We started with 3 horses and a John Deere model H. Today, many suburban gardeners have more power to work with. But the Deere could pull a one bottom plow 24 hours a day. We didn't just plow the wheat stubble. We plowed the bottoms and seeded them with bromegrass. We plowed the side hills into terraces to retain the water as far up the hill as possible. We planted 1,000 trees on the highest point of the property. The plan for the crops was rotation, fallow periods and alfalfa (deep rooted to survive transitory drought) and nitrogen fixing to restore the natural fertility of the soil.

The theory behind the trees was that they could alter airflow for 100 times their height possibly inducing reluctant clouds to yield their precipitation.

The wisdom behind the brome grass in the draws was not just erosion control, which it accomplished in striking fashion over a 5 year period by halving the distance from the tips of the hills to the bottom of the valley. It was also to slow or stop runaway water so it could be absorbed. In fact, in rockless light-soiled Nebraska we achieved a 6 foot waterfall between our farm and our neighbor's still ungrassed waterway. My father lumped a goodly portion of Dust Bowl wisdom into one sentence. "It's not as if we had a striking amount of difference of precipitation in the 30's, it is that the moisture fell— struck sun-baked earth and ran away before it could soak in."

The result of this thoughtfulness multiplied 10,000 or 1,000,000 times over was an outpouring of an abundance of crops nourishing post-war Europe and setting the stage for a sometimes uneasy postwar peace. That it was linked to the advent of the Machine Age for farming as well as the growth of center pivot irrigation systems made it appear an almost effortless American miracle.

You don't need a rerun of the vast surpluses or scandals involving too much crop production. You don't even need to know of the attempts to give the land back to the Indians or to the buffalo or to the Great American Desert except we are still searching for the Elephant in the Kitchen. This story may be useful at a later point.

Introduction to Dust

The legends of the dirty 30's reached far deeper into America than even I would have guessed. Recently I met a 60ish couple on a tour through the wilds of the Nevada Desert. For them it was a retirement celebration— not actual but retirement of the mortgage on their farm. Their farm was located 600 miles north of North Dakota. It was the section of ground that their grandfather had lost during the Dust Bowl depression. I was astonished to learn that they were greatly affected by the dust and drought of the Great Plains hundreds of miles north of North Dakota.

Now for part II of the search for the elephant. —dust!

The America of the 20's and 30's was an America of a rural population. When automobiles arrived in quantity the original speeds were relatively modest on mostly unpaved roads. As speeds increased the dust arose in greater and higher quantities. As tractor mechanization increased, the speed and quantity of plowing, teamed up with the famous drought to produce enormous amounts of dust that WAS CAPABLE OF ALTERING ENTIRE LANDSCAPES. It was also justly famed for seeping through invisible cracks in the woodwork.

We are mostly interested in the dust aloft. The pace of road and dam building increased as the decades crept toward 1940. 1939 saw the beginning of World War II but we are only interested in the dust potential of the awesome quantity of bombs dropped during WWII. Before the War was over we had discovered how to hurl enormous quantities of dust as high as a volcano and the A-Bomb was born. That dust making enterprise continued apace until 1962. Meanwhile, America's farm enterprise continued at a whirlwind pace. By 1958 the threat of European starvation diminished and the soil bank was introduced. Not very long after that "no till agriculture" was introduced. That meant a lot less dust in the air.

Now we continue with part II of the dust story—still looking for that great and mysterious beast— I'll assert that 1966 was THE pivotal year. The time is approximate but debatable. I'll document it with incidents from my memory. I started on evenings in June of 1966 for Yellow Freight System, Kansas City, MO. On a Friday night working on an open dock a dense fog crept onto the dock. Its density was so great that the other end of the dock was lost to view. Several of the men began coughing and an older man went into a paroxysm of coughing. "What is that stuff?' I asked a senior man. "The Aluminum Smelter next door cleans out their stacks before closing up for the weekend." With the benefit of age and good breeding I probably would have kept silent but I felt that lives were at stake here and I called the local TV station to complain. The American Alloys Plant was closed soon after but the relief was genuine and long-lasting because Yellow moved to its present airy location at I-435 and 70 Highways in Kansas City, MO. About that time the spokesperson for the automobile industry announced that they had found no earthly use in the human biochemistry for the megatonnage of lead that had once been the staple of gasoline octane. I have no realistic estimate of the tonnage of lead in the air or of its duration and so, no-lead gasoline eased its way gently into the American system. About the same time the biggest part of the Interstate system was completed which meant less construction dust in the air. The rural to urban migration was almost complete but the rural communities that remained were mostly served by hard surface roads, again, less dust thrust up into the winds.

Suddenly, everybody, particularly electric utilities became conscious of the dangers of particulate matter and filters of all shapes and sizes were included in almost every load of freight. Not coincidentally acid rain became a persistent and growing concern.

A whole body of changes then occurred in American Society from 1962 to 1980 and beyond to reduce and minimize dust and particulate release into the atmosphere.

The dust of the 30's was just dust. It imperiled livelihood and had great nuisance value as well as later anecdotal value. The dust of the 60's was of a different character entirely. We were warned of the hazards of lead, mercury, dioxin, DDT, tars, nicotine and cigarette smoke—even secondhand and then of bacteria, molds and viruses. Behind these obvious dangers lurked the spectre of radioactivity: uranium, plutonium, strontium 90, cesium 137, iodine, not to mention, cobalt and cadmium. It should not have been a surprise that the closing of American Alloys was accompanied by a ban on open burning, but the forgotten factor was the basic nature of some of the dust whose absence may have contributed to the acidizing of Americans diet, potentially inviting the growing scourge of cancer. *And only a few skilled authors have dared to link the acidizing of the oceans to the subsequent reduction in phytoplankton—Clive Cussler in ARCTIC DRIFT. P. 86. But none has linked it to the progressive diminishing of dust after the 1960's.*

In case you are lost I am describing a 35 to 45 year exposure to dust. It was followed by an equally dramatic recoil from dust and airborne particulates of many kinds. The relevant and dramatic comparison is to the overwhelming power of the eruption of Krakatau, Indonesia in 1883. Five years later enough dust had settled on the ocean triggering its own El Nino and Presto—the Great American Blizzard of 1888.

Revolt Against Dust—(Acidification?)

What the consequences of 35 years of dust could do to the climate we are spared from knowing because the 60's and 70's saw a veritable blizzard of books announcing the New Ice Age. One author even suggested that a single prehistoric volcano introduced the reading public to a New Ice Age that almost wiped out mankind. Nuclear Winter— a phrase attributed to Carl Sagan—entered our vocabulary. A fanciful movie was made to show that once it began to snow and the albedo (meaning reflective capacity) of the earth increased it could continue until the continents were completely covered with snow & ice.

Now before I embark on the next dramatic chapter of historical recharacterization, I need to remind you of two of the less well-known features associated with volcanic emissions.

We are all familiar with the particulate matter hurled into the air and blocking sunlight and providing us with gorgeous sunsets for a period of months or years. That cooling effect is rather direct and easily grasped. The first of these lesser known features is from the early fallout from the cloud. The dark colored ash from the volcano frequently floats and provides increased surface absorption of the sun's rays. This absorption is concentrated in the shallow waters, rather than penetrating to a depth of 200 meters or more. The differential heating of slightly murky water can be easily illustrated by a summer time dip into a moss or algae covered farm pond. The top 12 inches is too hot, the next 5 feet is cold. Considering that water tends to vaporize at or above 84 degrees, this top 1 foot or so of murkiness acts to produce a sharp increase in the rate of evaporation. The second of the lesser known consequences of volcanism is supplied by the remaining airborne—lightweight—dust particles. They provide plentiful nuclei for cloud formation.

This moves us to an additional ingredient in comprehending volcanic influence on climate—the invading confrontation with a cold air mass. This is so important that we'll spend more time on this in a later chapter. Lastly, and to be amplified later, the onshore site of the resulting rainfall can and should be high enough and receptive enough to slow its return to the ocean so that re-raining can take place. The engineered hardness of the ground—paved streets-parking lots-roofs—coupled with the natural sun-baked hardness or the endless sand with the root systems baked out of it provide a considerable hurdle. The picture below begins to illustrate the ideal depth of The properly rain-receptive soil!

DEEP
RICH
MOIST
BLACK

Picture redrawn by Amber Leathers

Before we depart this section with its lengthy comparison with the five year influence of Krakatoa, **let me re-emphasize another five year cause and effect sequence. The nine day wonder (oil spill) in Lake Maracaibo in 1922 was followed five years later by the all time greatest 1927 flood of the Mississippi.** No meteorologist that I know of has linked the two events in a causal sequence so I'll pass it off with the nation's meteorologist comment, "The last 50 or 100 years have been characterized by abnormal weather." Coming of the New Ice Age p.__?_. Meaning," it has been unusually nice." The prediction was for a return to planetary normal. Meaning "please be nice to us forecasters because we have no idea of what to expect."

Was the forgoing a *flight of fancy?!* I think that I have documented substantial delays to the evaporation process caused by substantial oil spills enough to recommend extensive remedial aids to global water evaporation. But it lays the foundation for the *Fact Section*, establishing the work area!

Chapter 3

Thesis: from the Ency. Britannica 1966 edition vol. 22 pg. 338; "**SOCIETIES OR CIVILIZATIONS** (not nations or periods) are the significant units of historical study. 'Civilizations, 26 of which he distinguishes, including arrested societies, grow by responding successfully to challenges under the leadership of creative minorities; they decline when the leaders fail to respond creatively." Toynbee

Now I'm ready to tackle the next step in this little drama—I hope you are. This is one more perhaps myopic glimpse of the elephant—please don't leap to conclusions. If I had been intensely aware of the weather in 1978 which I was—and extended that awareness to the climate—which I did not at that time. And if I had power—which I do not—and devoutly hope to not possess. And if I had reason to believe that a new ice age was imminent, I might have sought to slow the evaporation of the oceans which would have slowed the fundamental foundations to the overwhelming onslaught of precipitation which in winter time could provide the snow base to initiate the necessary extra albedo.—*Presto* . . . *instant ice age* . . .

Preview To The Relief From Water

That's way too many if's. Nobody could build castles in the air or snow or sand that high. But according to the available record that's exactly what happened. *The 2001 World Almanac and book of facts*, details five incidents: two blowouts, a collision, a fire and a grounding that occurred in a span of five years beginning Mar. 16, 1978—culminating in a blowout in the Nowruz oil field in 1983 in the Persian Gulf. These spills were estimated at a grand total of 1,973,000 tons of oil. To place this astounding figure into perspective I compiled the tonnage of oil spilled from the 1989 Exxon Valdez spill until 2002 at 713,000 tons roughly 1/3rd the volume, and almost 3 times the time period. To achieve this statistical comparison I had to omit the Persian Gulf fiasco out of consideration but only for a time. That spill deserves its own chapter.

These figures come from the Internet, *http://www.endgame.org/ oilspills.htm* 8/29/2007. These, are courtesy of George Draffin but any errors in compiling the approximate tonnage are mine. See accompanying table and map. Pages 26 and 27.

If stupefaction is your goal you can stop reading right here. You are wrong again if you believe that I believe that I've described the elephant. The correct reaction is intense admiration for the originators of the spills. That reaction is of course correct only if you believed that we were headed inevitably toward the next ice age. That question is reserved to a later more extensive chapter titled: "Ice Age—To be or not to be?" The rabbit trails of thoughts about these spills are intended to be laid to one side by the prologue to this chapter. In other words, societies are challenged, minorities respond, creatively we hope. Before we get to the brutal, dessicatingly hot facts of the work area, please allow me an apparent digression, which in reality, is a

personal introduction to the work area

In the eighties I had no immediate response. Six children, a double hernia and 20% inflation successfully kept my nose to the grindstone. But I did have a response by the late eighties, and it may have substantial validity even today. I'll tell it in story form with some of the dates approximate.

In the late eighties or early nineties before the floods of '93, my wife took note of my rundown condition and asked me if there was any special place that I would like to go. I already had a place in mind. "The inland passageway to Alaska is supposed to have some of the most beautiful and spectacular scenery on earth." She secured two berths aboard the largest passenger liner in the world. The trip was sponsored by the John McArthur Ministry.

We boarded in Vancouver and set sail. The opening festivities included a planned social hour—ostensibly to get acquainted with some of the other passengers and begin to feel at home. It was my lot to be placed at a table with an older gentleman. No others sat with us. The conversation began with his opening statement, "I'm a retired farmer from Bakersfield, CA." I must have sounded miffed with my reply, "I'm from Nebraska and Missouri" and maybe I added another line sounding like, "that's real farm country." At any rate the reaction was similar to waving a red flag in front of a bull.

His chest swelled and he began, "If my county were a state and California was left out—my county would be number 1, Texas would be number 2, Iowa would be number 3, and another county in CA would be number 4 in the value of agriculture. "Goodness" I said in an appeasing tone. "How big is your county?" His answer, "70 by 120." I thought to myself but didn't say, "Nebraska is about 10 times that size and 93 percent arable." China has a billion people and their land is 13 percent arable. But I said only, "You have to raise a lot of crops on that amount of land. How many?" His answer has escaped me because I later compared it with Israel's claimed 6 or 7 crops and self-sufficiency in 2001. His answer also was 6 or 7 and I apologize if I have offended his ego by my poor memory but by comparison—Nebraska at that time was fortunate if they could get three cuttings of hay. But I said only "you have to have water for that!"

The effect was astonishing—almost like sticking a pin into a large balloon. His puffed up chest collapsed because, California was in a six year water shortage. "Let me spin you a water story" I said, "I want to dig a water carrying tunnel—the third or fourth largest river in America—from Los Angeles basin to Death Valley California. After I have installed electrical generating equipment to harvest the power of the water falling two hundred feet into the basin, I will install mirrors to augment the evaporation rate so that water can continue to flow unimpeded. The lake will become a laboratory for biological remediation—oil removal and conversion to agriculture—carbon sequestration to the sophisticated. The heated saline water will return to the Pacific where it will eventually meander to the surface again closer to the North Pole. That heat surfacing into cold air should contribute vitally needed wintertime precipitation to the melting Artic.

That extra water evaporating continuously through the torrid California summer and perhaps re-raining on the Rocky Mountain slopes to the East might provide an important building block to increase the summertime humidity so that the Santa Ana winds and their enduring hazard to L.A. might be minimized or eliminated." (More solutions for L.A. will appear under Australian water harvesting techniques.) [The 260 gallon rain barrels that they used to trap the sparse rainfall would not be sufficient for the size of California housing. A more appropriate size to harvest California's winter rains would be 1,000 gallon tanks on each corner of the house, that would be released at precise summertime intervals [AND, perhaps also some calculated January morning release of the abundance of wintertime rains in order that the sunshine could augment the west to east flow of moist Pacific air and restore a normal snowpack on the heights and slopes of the Rockies, to bring a measure of control to California's epidemic of dry season forest fires.]

After this monologue I should have stopped, but I added this sentence. "When I am finished I intend to sell the most expensive building lots on earth because I will have created a Hawaiian style climate on several sides of the lake." It was probably overkill because I had his fullest attention. I leaned forward and added, "You probably think I'm doing this so that the extra mountain precipitation will drain in the direction of your farm in Bakersfield. No!" I said. "I'm not even doing it so that the Colorado River will once again run to the Ocean. The reason I'm doing it is so that extra wintertime precipitation will resupply the Ogallala aquifer on the East side of the Rockies where the real farmers are ready to use it."

Needless to say the session was over. Neither of us said boo to each other as we parted. The week of the tour was beautiful. Meals were abundant and plentiful. We stopped at a salmon hatchery where the guide informed us that of 55 million hatchlings released only 5 million had returned. It was about this time that the undersea ocean temperature was two degrees above normal and that fish were being caught as much as 2,000 miles north of their usual haunts. The guide hinted that a new breed of salmon was considered because this one was probably worn out.

We got close to one of the receding glaciers, Mendenhall I believe and then turned back toward Vancouver. The trip back was uneventful with a few pleasant strolls on the upper deck of the tallest passenger liner in the world. Dolphins and whales displayed themselves before our cameras. And my thoughts turned toward the inevitable return to the world of work. I had carried our bags to the disembarkation point and was sitting on them when I fell into conversation with an old gentleman who was also ready to disembark.

Our conversation was one of those once in a lifetime of camaraderie. For once in my life I could say nothing wrong. Good ol' boys rarely sustain the appearance of amicability for so long. After we had covered about 10 or 15 topics without finding anything to disagree about he suddenly stopped in mid-sentence. "You don't know who I am, do you?" "Your face looks familiar." I said. He replied with a touch of asperity. "I'm the retired farmer from Bakersfield." "Oh!" I said relieved to remember him. "You're the guy I spun the water story to." "Yeah," he growled and added in deep bass. "And if you ever get serious about it, I'll write you a BIG check."

When I tell this story, it happens at least twice a year for the last fifteen years, people almost invariably laugh before I get to the punch line. Here it is, "I still don't know that farmer's name or address." But I'm still serious about the project. The pragmatic update for the 'manufactured' sea in Death Valley, provides for two freshwater lakes on either end of the artificially created 'sea'. I don't remember whom I should credit for this welcome addition.

A Common Sense Bridge to the Work Areas

The reasons for being serious can be lumped under one catch-all term, *rising ocean levels*. But that term obscures the multiple factors involved in the ocean surge. Let me conclude one part of this oily mess by noting that some atmospheric scientists estimate that 2 to 3 percent of the earth's atmosphere is the most potent greenhouse gas: water vapor. Earlier estimates ranged from 1%-4% Not surprisingly, I read a more recent estimate of 1%-2% water vapor in the atmosphere. One estimate even ranged from 0 to 2%.

Intro To Rising Ocean Hysteria

If that 3% water vapor were to suddenly plop out of the atmosphere at once it would amount to 12 inches of rain precipitation on every square foot of the globe. That 2% or 3% estimate is figured in this way. The weight of the atmosphere is capable of raising a column of water 34 feet into the air. 3% of 34 feet = 12 inches, there are other estimates, but for now I choose to illustrate with these figures. Do the figures yourself, don't trust me! But to continue this seemingly idle vein of thought, if the proportion of atmospheric water vapor were reduced from 3 to 2 percent by the surge of equatorial oil spills during the late 70's and early 80's that could explain almost the entire estimated ocean surge.

To my limited knowledge no respectable meteorologist or computer modeler has factored this into the endless stream of possible disasters culminating in the Hollywood blockbuster, "Water World."

Let's conclude this personal interjection by agreeing that these two old men, one farmer and one "wanna-be farmer" found common cause in noticing that the world around them had altered its water supply character. The irony is that in 1980 Nebraska corn was the only commodity that was not subject to the universal law of inflation. In fact its price hovered for years some 20% lower than its price in 1948. Only in (THE LATE 90'S?) with oil selling between 60 and 70 dollars per barrel did we realize that an equivalent weight of corn (energy) could be purchased for less than 12 dollars.

This oily mess is about to depress me. But before I switch to a more fact-filled argument let's briefly note that the relatively minor creek in Nebraska that runs not far from our hill-top farm recently was awarded 6 (count 'em), 6 flood impoundment reservoirs. That award takes me back in time to a 1947 Reader's Digest article. That article took note of the extensive progress made during and after the 30's in building large dams and power generation facilities. The title was "Big Dam Foolishness!" It didn't quarrel seriously with the big dams, but it made the very valid point that small dams further upstream were vitally important to recharge soil moisture and alleviate the boom and bust cycle of Great Plains agriculture.

Further guidance for water hungry societies: Nebraska's rainfall in most years qualifies it as a desert. In a world where rainfall is irregular, water impoundment is a must to maintain soil in a rain receptive condition. Which brings us to the semi-political point which must be made decisively. A world which irregularly anoints its oceans in a haphazard accidental way must transfer its primary value from oil to *"CLEAN WATER!" It is for this reason that I gladly accept and welcome the very able help of Al Gore in this complex and emotionally entangling desert maze!*

Intro To Center Pivot and Potential For Sahara

This does not presuppose an undying conflict with "oil" as in "oil and water don't mix." Nor does it have to detract from the inherent value in oil powered energy. Witness the example of the Saudi's! On the eve of the mideast conflict I selected an up-to-date atlas of the world. It included an aerial satellite picture of Saudi Arabia. I counted over 1,300 center pivot irrigation systems. The great desert was liberally spotted with green circles. A near perfect marriage of oil and water and not only in Arabia; the good ol' Great American Desert Nebraska is 63% served by center pivot irrigation systems.

And when the logical extension of this idea receives the factoid from page 71 of the Reader's Digest, "Strange Stories Amazing Facts "(1976) to wit, "the Sahara Desert has an underground reservoir of some 150,000 cubic miles." One side of the rehab job is made apparent to us. It is accompanied by this question. Given the 1976 date of the estimate and the excess oil releases from 1978 to 1983, how much of that 150,000 cubic miles was diminished in the next 15-20 years?

The Second Work Area—Afar 3,000,000 square miles of desert sand ought to qualify readily as a filtration model for clean water. The next best source is evaporated water. One prime area is identified in Scientific American Oct 08 pgs. 60-67. The "Afar Depression marks the north end of the East African Rift, a 3,500 kilometer-long zone of tectonic turmoil that is tearing the continent in two . . ." Article titled **"Birth of an Ocean, " story and photographs by Eitan Haddok"** Pictured are drifted deposits of salt. The caption: "Ghostly salt deposits near Afdera volcano testify to ancient inundations in Ethiopia's Afar region. In the past 200,000 years, the Red Sea flooded Afar's lowlands at least three times; the salt stayed behind as the seawater evaporated. One day the ersatz seascape will likely become the real thing . . .

"In northeast Ethiopia one of the earth's driest deserts is making way for a new ocean. This region of the African continent, known to geologists as the Afar Depression, is pulling apart in two directions-a process that is gradually thinning the earth's rocky outer skin. The continental crust under Afar is a mere 20 kilometers from top to bottom, less than half its original thickness, and parts of the area are over 100 meters below sea level. Low hills to the east **are all that stops the Red Sea from encroaching."** . . ."**for now,** this incipient seabed is a less desolate landscape where lava stifles vegetation, hellish heat makes acid boil, devilish formations emit toxic fumes, and the salty legacy of ancient Red Sea floods provides nomadic tribes of Afar with a precious export."

When we allow the (electronic or rapid fire) [even print] media to shape the message, it forces an unconscionable narrowing of decisions for remediation. Here the context is 'rising oceans'. Media hype and satellite cameras have found an easy scapegoat in the pictures of 'disappearing' Arctic and Antarctic {sea} ice. If we can trust the statement that this area has been flooded 'at least three times' in the last 200,000 years, we are face to face with the least expensive alternative to 'rising sea levels inundating our coastal cities.' The engineering problems involved in introducing Red sea water to that hotbed of solar and geothermal heat could perhaps be solved in time to avert the next extreme of storm surge which produced an estimated $20 billion ("Hurricane Katrina . . . caused damage estimated to be as high as $156 billion." [Guinness World Records, 2011]) price tag for **some** aspects of the New Orleans 'sinking.'

Work Areas By The Numbers

And now for the continuation of the search for the hidden quadruped: the search for clean water. We are starting with facts from the 1965 edition Encyclopedia Britannica volume 12, page 6 and 7. Concerning the Imperial Valley and the Salton Sea, "The entire valley is below sea level, barely below at the Mexican border and 235 feet below at the edge of the Salton Sea. Were it not for the broad, dam like deposits of the delta fan of the Colorado River, the waters of the Gulf of California would invade and drown the Imperial Valley, the Salton Sea basin and much of the Coachella valley. The Imperial Valley has a hot desert climate, similar in many respects to the Sahara in Africa. The combination of high temperatures, abundant sunshine and numerous strong and drying winds makes the scant rainfall extremely ineffective. The lake level varied between 195 and 234 feet below sea level."

This is Egypt, Encyclopedia Britannica volume 8 pages 28 and 29. "Western and southern deserts, this region covers an area of 260,000 square miles or nearly 3/4 of the whole country (Egypt) apart from a very narrow coastal strip of gently undulating plain where some rain falls in winter, it is one of the most arid regions of the world. The ground continues to fall toward the Mediterranean and to the depressions of the Siwah oases and Qattarah, the latter being the largest containing lakes and marshes and descending to 435 feet below sea level. These depressions, except Qattarah, are habitable because of artesian supplies of water."

Encyclopedia Britannica volume 7, page 136, "Death Valley, in Inyo County California, is lower, hotter, and drier than any other area in the United States. It is bound on the east by the Grapevine, Funeral and Black Mountains and on the west by the Panamint Range. Death Valley is in many ways unique. Some 550 square miles of the valley floor lie below sea level, culminating at Bad water (-282 feet)."

I'm saving the granddaddy of all for last but I'm not quite ready to reveal my biases or at least the two or three that I know about, until a later chapter.

From the Encyclopedia Britannica volume 7, pages 116 and 117. (1966 edition), "Dead Sea, the lake into which the Jordan River (q.v.) flows. Its northern half belongs to Jordan; its southern half is divided between Jordan and Israel. The Jordan Valley bounds the Dead Sea on the north, where part of the shore line consists of deltaic salt marsh. There the valley is 12 miles wide, with badlands and scrub bordering the delta. It is bounded on the east by the fault escarpment of the flat Moab plateau. The escarpment rises 3,000 feet above sea level, an average ascent from the lake of 14 feet per 100 feet of horizontal distance. It is bounded on the west by the rolling surface of the Judean Mountains and northern Negev highlands, where peaks rise to 2,300 feet above sea level. The Dead Sea has a maximum length of 51 miles, a width of 11 miles, and an average of 1,080 feet for its deeper portions and a surface area of 394 square miles 3/4 of which is Jordanian territory. It's surface level, which has a seasonal variation of from 10 to 15 feet, lies 1,302 feet below the level of the Mediterranean Sea and is the lowest sheet of water on the earth's surface." "(It is considerably deeper & saltier now than the 1966 level)."

It goes on to say, "In the summer the absence of rain and the high rate of evaporation cause water levels to drop from 10 to 15 feet below those of winter. For centuries the level of the lake rose, because of increased rain, increased intake in relation to evaporation, depositional fill, increased flow of subterranean springs or combinations of such factors. From 1879 to 1929 however, the water level rose only 23 feet, and since then it has dropped a few feet. There is no commonly accepted explanation for this phenomenon." *{It couldn't be that the massive oil spills half a world away in the Caribbean affected the amount of moisture available to be milked out by the slopes on the east side of the Dead Sea. Surely I'm not the only Horner in the world to come up with this theory}.*

We continue the encyclopedia statistic, "Intake from the Jordan River is 6,500,000 tons daily. Other streams numerous springs and underground seepage add another 500,000 tons. Four perennial streams; Wadis Hasa, Mawjib, Zerqa and 'Udhemi flow into the lake. They all enter from the Moab plateau, which receives approximately 16 inches of rain in the winter from moisture bearing winds that rise over the steep western facing escarpment. The intermittent streams flowing into the lake from the west, originate on the leeward side of the Judean plateau which receives from 4 to 8 inches of rain in the winter. Rainfall in the Dead Sea valley averages 3.4 inches annually in the north and 1.8 inches in the south. Seldom does it exceed 5 inches anywhere in the valley. There is no outflow from the lake; instead the water balance is maintained by evaporation, estimated at 120 inches annually. Blue white clouds which form a mist over the surface of the water carry off this evaporated moisture.

"SALINITY—Dead Sea water is intensely saline, with solid content of 25% as compared with from 4% to 6% for ocean water. Dead sea salts include magnesium chloride, sodium chloride, calcium chloride, potassium chloride, magnesium bromide and calcium sulfate. At the surface, each litre of Dead Sea water contains 227—275 g. of dissolved salts. At a depth of 360 feet, the water has 327 g. of salts. The high percentage of solids in solution makes the Dead Sea extremely buoyant, keeping bathers continually afloat. Fish are not able to live in its waters."

Well! What do we have here besides ancient Encyclopedia data? For starters, we have the four hottest and lowest and driest places on earth. Additionally, they have exaggerated drying capacities courtesy of the strong south winds and the sunbowl-like surroundings. Additionally, they are all far below sea level. If non-petroleum (carbon based) power is your goal, these are wonderful candidates. That is not even close to an adequate reason but a reasonable case could be made for these projects by a good CPA. In fact, Israel is doing a feasibility study—more on that later. Two of them have excellent access to water. The Qattarah depression is practically on the prehistoric site of the West Nile. We have already mentioned that the Imperial Valley is dangerously close to being inundated by the waters of Baja California.

I have tried in my homespun way to prepare you for the shock of my intentions. Let me review briefly before I take you further along this path. We have seen two farmers note the change in water availability. This shortage was marked on one end by the super heat of 1980. It was marked on the other end by the equally dramatic floods of 1993. In between were desert-like extremes of hot and cold punctuated by the destructive Florida orange groves frost. America hardly noticed this world-wide crisis because the Ogallala aquifer supplied a more than adequate insurance policy, via the advance in center pivot irrigation systems.

Oil Drought As Motivation

The Oil *bleepity bleep* Drought? Whatever it was lasted almost 13 years. The old rules for drought, 7 years, established informally during the time of the Biblical Joseph's saga, slavery, exile, bondage, jail and finally helpful creative rule in Egypt. These rules were not exceeded during the time of Elijah 3 ½ years by the reckoning of most scholars. Even during the Dust Bowl, the rules of seven year droughts were paid at least lip service by the many scholars who studied and analyzed the phenomenon. The actual term of the drought varied with the area of the country and was complicated by the length of time that the Depression extended our public perception of it.

If 13 years seems long to you, save some of your sympathy for Israel. My wife and I were granted the luxury of a two week tour there in early 2002 after retirement from Yellow Freight. There, we saw a sophisticated miserly system of drip irrigation. Our guide announced that Israel had achieved a positive balance in agriculture. But on the other side of the coin, the sea (lake) of Galilee was more than half depleted and a net thrown over the side during our boat ride produced only an empty net when drawn back into the boat. On venturing to the Dead Sea, our fact sheet proclaimed a salinity of 24% but our guide assured us that what we were looking at was two lakes; the north had salinity of 27% and the south end a horrific salinity of 33% with mounds of crystalline salt protruding at several points in the lake. The information from that site was coupled with his proud announcement at Galilee to the effect that Israeli divers had gone to the bottom of Galilee and either capped or diverted several salt springs in order to make the water fit for human use. This was done earlier in Israel's post 1948 history.

Even though I have almost 10 years exposure to western Asian and southern Russian exchange students, I don't have enough knowledge at my fingertips to draw a straight timeline connection from the massive blowout of 600,000 tons of the Nowruz oil field in February of 1983 to February of 2002 in Galilee and the Dead Sea. It is safe to say neither that spill nor the later deliberate release of oil (estimates vary wildly-World Almanac 2001-130,000,000 gallons others are high as 460,000,000 gallons) didn't do any of the western Asian and southern Russian countries any lasting good—YET.

It seems to me not an impossible leap to note the two massive Iraqi oil spills of '83 and '90 and wonder if there might be a causative connection (among many others).). *In an oblique speculative aside, It has occurred to me to wonder if the '83 spill was triggered by a clever and desperate Saddam Hussein who might have known that West to East winds would deprive Iran of life-giving rain in the middle of his desperate 8 year war. The only factual confirmation for this bizarre theory is the unusually strong insistence of the Iranians on achieving atomic power. But their motivation could also have been derived from the equally false assertion that "fossil fuels are precious and about to run out. (more to this point later).*

The usual attribution is to the excessive use of irrigation water by the countries surrounding the Jordan River.

It is my considered opinion, which I share with Dr. Toynbee that 18 years of water hardship for Western Asians could spur the development of a creative minority to respond to the enormity of this challenge. It need not appear so vastly different from the response in America to the disastrous 1927 Mississippi Flood. That flood promoted the Army Corps of Engineers to take charge of a wide variety of successful flood control and conservation projects throughout the central and Western United States.

Water Adaptation USA, Saudi & India Now

But they don't need to take a penny's worth of advice from the United States. The examples of Israel and Saudi Arabia are plentifully replete with wisdom and well-used wealth. Here is an example from recent travel. While standing in line for belated boarding of a plane, I struck up a conversation with a young Indian (India) lady. She was gainfully employed as a coordinating supervisor of some international agencies dealing with health issues such as AIDS. This young lady, had she been American of that age, would have been qualified to converse about public health and medication. But our small talk introduced me to a brand new phrase—rain harvesting. The India of Gandhi and Nehru and the Ganges and the Brahmaputra suddenly focused on, of all things, rain harvesting. This next material adds some reasonable flesh to the conversation with the rain harvesting Princess in 2002". Source: page 137 of Herman Kahn "The Next 200 Years" publication 1976. "India is blessed with extensive and fertile river basins." The need for irrigation and development of water resources is one key to greater production since monsoon rainfalls occur only over a 4 month period, leaving a large part of the country semi-arid for the remainder of the year. The development of Ganges basin (alone) could increase grain production by 150 million metric tons or more.

Aside from Herman Kahn, my contact with the young woman was no professional survey, but if water consciousness was raised to such a level that a young professional woman could converse intelligently about water, it suggests to me that Saddam Hussein's manic releases of oil could achieve for him in death what he could never achieve in life with all of the military hardware that he could beg, buy, borrow or steal. This could potentially unite all of the water starved areas of Western Asia into a governing entity with its main goal to provide for adequate clean water for one of the most water-short populated areas of the world.

Water Adaptation for Future Israel and Jordan and Palestinians

Oh! Speaking of water unification, "On May 9, 2005, Jordan, Israel and the Palestinian Authority signed an agreement to begin feasibility studies on the project, to be officially known as the "Two Seas Canal". This scheme calls for the production of 870 million cubic meters of fresh water per year and 550 megawatts of electricity. The World Bank is supportive of the project" (Secondary reference, original lost) *http://www.2.dcn,org1pipermail1enu-trinity120071001293.html I have a salutary rehabilitative reason for supporting a Mediterranean to Dead Sea Canal. That argument will be illuminated by the need to restore the evaporative preeminence of the Atlantic Ocean and its potential benefits to world civilization.*

But, we aren't done trying to catch a glimpse of the elusive multi-colored elephant. Saddam's spills could qualify him as a user of weapons of mass destruction in the eyes of the suffering nations around him, but the use of that language would have elevated water politics onto the political table. Which is about where it should be!!

I hope that I have convinced you that real change is possible and that the work areas are well chosen. I will outline additional applications for Death Valley that may be accessed on an emergency basis with a lot of work, but it should be done only as an emergency part of a coherent plan.

Waterless Adaptation of Past Saharan Sand via Richie Calder

Here we can take a peek at history through a new set of eyes. Ritchie Calder, a journalist wrote a book titled "After the Seventh Day—The World that Man Created." At the time of my first reading, mid 1960's I suppose, his major thesis that the Sahara Desert, lush grassland several thousand years ago, was converted to a sandpile largely by the hand of man. He documents the wasteful behavior of the ruler of NE Africa, the building and sinking of thousands of ships in the Mediterranean Sea, and the destruction of the Cedars of Lebanon in the building of Solomon's temple . . . as if we needed proof that there was once abundant vegetation; He retells the tale of Hannibal's trek across N. Africa across Gibraltar to defeat the Romans using elephants in 218 BC.

Scientists long suspected that the Sahara wasn't always dry. Then, during a 1981 space shuttle mission, NASA used a synthetic aperture radar to peer through sands to the desert's underbelly, revealing a network of river courses deposited in the bedrock during multiple wet intervals.

Courtesy of NASA

How much more evidence do we need, before we can really believe that the Sahara was once wetter and more than adequately vegetated.

My father's work and words of wisdom on the farm had prepared me to believe almost before I read the book. The additional line to the effect that once the sand is exposed and heated the rain can fall but cannot reach the ground because it evaporates before it reaches the ground, the applicable word is virga. The mass of heated air rising from the sandpile then becomes the engine driving the jet stream and thus future rainfall further north.

By the way, it was my Nebraska in-laws, who reminded me of the correct spelling and usage of the term VIRGA—not my beloved sis of NW N. Mexico with 30 years of experience around the Methodist Mission in Shiprock N. Mexico. I can't quite see an army of elephants being fed today from N. African vegetation.

Lest we miss a really easy opportunity to use the symbolism of the elephant; I need to remind you that the elephant in this case is fixed in the public mind when the word "desert" is spoken. We have images of endless sand, desiccating heat, impregnable fortresses, oh yes, and Omar Sharif.

We need an alternate symbolism here. The desert southwest in America has two of the fastest growing cities in the world, Phoenix and Las Vegas. Even Los Angeles is at least a partly planned city around the entrapment of water. There will be more Planning for LA to come later.

Let's begin to alter the symbolism of the Sahara with this story taken from the September 1999 issue of the National Geographic. The grandson of the famous Auguste Piccard Bertrand attempted to duplicate the round the world trip by balloon. He found himself chopping two ten foot icicles from his balloon at the height of 23,000 feet over Mali, the heart of the Sahara. That's visible proof of the meteorologists bold statement, "there's plenty of water passing over the Sahara Desert, all you need to do is make it fall all the way to earth!" If this does not put a glimmer of an idea in your head Sir Richard it is because you like me have lost track of the number of stationary satellites whirling around earth.

To help the idea take root; let me share with you a largely apocryphal story of Florida in the decade of the 80's. It seems that they had a summer so dry that a meteorologist came up with the descriptive phrase, "once in 500 years drought". It was a catchy number and apparently resolved public concern about the appalling heat. Then the next summer came along as bad or worse so they had to dig a little deeper in the bag of reasons. Being unwilling to mention oil and Florida Sand in the same breath—might be bad for the tourist image—they blamed it on the farming and development practices which removed trees and drained swamps, thus allowing the sandy soil to heat up and courtesy of virga—presto, instant mini-Sahara. In this case they were in my farm-aided judgment more than half right.

Greening the Sahara may be a bit more complex than cooling the sand; but the miracle of green can't begin without water. If we don't set our aims high enough we wouldn't recognize a miracle if it hit us in the face.

Chapter 4

ONE MILLION YEAR ICE AGE CAUSED(?) (BY AFRICA?)

B efore we leave the subject of the Sahara and move to potentially greener pastures; did you know that we have been locked into one Ice Age for the last million years? I was taught differently in my country school, high school, and college. The shock of disbelief was suspended temporarily by the August authority of the American Academy of Science placed around the revelation. The approximate words were, "one ice age with periodic recurrences for the last million years based upon the present alignment of the continents." Since I'm not going to contribute to changing the location of the continents; it remains for me to figure out how the location of the continents could contribute to a million year Ice age.

The best candidate that I have found is Africa specifically the intensely reflective northern half called the Sahara. For reinforcement of that idea I turn to a more recent study of the Mediterranean Sea document title Circulation and mixing of Mediterranean water west of the Iberian Peninsula. Essentially it documents the diffusion of hot saline currents exiting the Strait of Gibraltar into the Eastern Atlantic, (Internet) http://

cat.inist.fr/?aModele≠afficheN&cpsidt=3552483. An earlier document claiming the American Academy of Science backing claimed that a significant source of Polar warming was provided by exiting hot saline water from the Mediterranean.

Much earlier in pre-history the legend of Snowball Earth is supposed to have its origin at a time when the continents were mainly located along the equator. Again, the theory is that land being much more reflective than water could contribute to a massive lessening of gathered and stored solar energy.

Suffice for now to note that the Sahara is a prime candidate for carbon sequestration, the more so because mature forests frequently give off as much CO_2 as they absorb. Additionally, the oceans have exhibited occasional signs that they do not have unlimited absorption capacity.

Charles Mallory Hatfield—Con Man or Ahead of His Time?

Lest you are ready to emit a despairing sigh and pronounce me to be another useless Don Quixote, let me share with you a story that is well attested and unquestionably true as well as a bit of a mood lightener. From "Freaks of the Storm" by Randy Cervany, "the world's strangest true weather stories", page 140 ff.

"Undoubtedly, the supreme rainmaker of the early twentieth century was Charles Mallory Hatfield of San Diego, California. According to the detailed expose given by Clark Spence in his 1980 book the Rainmakers, Hatfield started his adult life as a sewing machine salesman but quit that job in 1902 to begin his illustrious career as a rainmaker. As his initial job, he accepted a commission of $50 from a farmer to "precipitate moisture' near Los Angeles in 1903. Reports are that Hatfield built a wooden platform that stood about 20 feet off the ground for his rainmaking experiments. On that platform he then placed large holding tanks in which he mixed a secret set of chemicals that he claimed would induce rain. Apparently Hatfield's work in that initial job succeeded beyond anyone's expectations. The farmer reportedly even paid Hatfield an extra $50 in appreciation for the rain that Hatfield 'produced'.

After that opening success, the Great Precipitator as Hatfield came to be called conducted more than 500 rainmaking contracts in and around Los Angeles over a period of 25 years with fees steadily increasing. Within a couple of years of his start, Hatfield took on a four-thousand dollar commission to fill Lake Hemet reservoir; promising four inches and delivering more than seven. His chemical releases were followed by eleven inches of rain that raised the water level in the reservoir by 22 feet and according to the testimony of the operators, gave them the biggest bargain they ever had. Popularity for the rainmaker began to swell. Merchants even began marketing 'genuine "Hatfield" Umbrellas' when they heard the Precipitator was coming to their area.

Hatfield's greatest and most infamous rainmaking adventure came in 1916 when he and his brother Paul were commissioned by the San Diego City Council to end a drought in Southern California and fill the Morena and Otay reservoirs. San Diego's contract with the rainmaker stated that Hatfield would receive a fee of $10,000 if he succeeded in producing rain to fill the Morena reservoir within one year but the Great Precipitator would receive nothing if he failed. Nevertheless, the Hatfield's set up their 20 foot platform near the reservoir and began a series of rainmaking procedures. On January 1, 1916, Hatfield launched his first rocket of hydrogen and zinc smoke. Nine days later rain began to fall in San Diego. And it kept raining. By January 19, the Morena reservoir was filled to its capacity for the first time since its construction and it continued raining. Water began to spill over the top of the dam; highway bridges and railroad tracks were washed out; telephone and telegraph lines were disrupted, and it kept raining.

The San Diego River overflowed its banks. Thousands of people had to be evacuated from flooded areas of San Diego and still the rain fell. On January 26, a record daily rain-fall of 2.41 inches occurred over the San Diego area. Both the Moreno and Otay dams ruptured and at least 12 people, maybe as many as 50 were killed. Property damage in the millions of dollars was recorded. By this time, the surging floodwaters had turned the tide of public opinion strongly against the great rainmaker. A front page cartoon in the San Diego Union depicted a mad farmer chasing Hatfield into the bay. Rumors had it that lynch mobs combed the city for Hatfield in every hotel and boarding house.

The San Diego City Council's reaction was even more disappointing to Hatfield. They flatly refused to pay the $10,000 fee they had agreed upon even though he had quite obviously fulfilled his part of the contract. After all, the Moreno reservoir had been filled. But, according to the councilmen, they had hired Hatfield to save San Diego not to submerge it. So they offered Hatfield the practical choice of assuming legal and financial responsibility for the inevitable lawsuits from the flooding, or leaving town with no pay and declaring that the rain was 'an act of God'. Hatfield wisely chose the latter option.

That incident was not so light-hearted back then but in a world where rising oceans threaten massive death tolls from coastal flooding it assumes a different dimension. I became personally acquainted with a young man (old now) who participated on at least two seeding experiments with similar success to Mr. Hatfield's. The project was dismantled because of overabundance of results not because of hocus-pocus and lack of results.

The story above is then . . . The picture below is now! How we got from one to another is part of the story of the book. The remainder is how to get at least part way back, to water adequacy.

Constipation In The Eastern Pacific—Picture

Amber D Leathers

Now that I've begun the process we should explore some of the overwhelming natural forces which can threaten or overturn the best laid plans of 'mice & men'! The foregoing seeding experiments might place a yellow caution light on some of our choices. But out there in the real world we have "necessary constraints" imposed by the certainty that we have about as much influence as a fly upon the wall on the real climatic, continental, electromagnetic as well as cosmic, volcanic, and orbital causes of Ice Ages. But insofar as we begin to understand our former role as a catalyst for change especially harmful change; we can then cheerfully and ethically shoulder the burden of righting those wrongs.

Handicaps to Real Planning

I neglected to mention the power of sunlight in the previous list of climatic influences that may be beyond our power to alter because sunlight is already a variable as it leaves the sun, but its power as it comes through the atmosphere is a proven variable. Witness the ozone hole, witness the man-caused epidemic of dust from the Late 30's to the mid 60's. Certainly I had a cloistered upbringing, but I had just barely heard (not seen) of skin cancer before 1960. Real science to me would document some of the increase of ultraviolet light near the equator and its crippling effect on ocean vegetation. The only statistical marker shows in the 2 year decline in the rate of CO_2 growth at the peak of dust fallout of 1964-1966. Some real Science has belatedly crept in from the Science channel 8:21 '08?, 9:00 a.m. Lightning strikes contribute ozone both up and down during a thunderstorm and the relevant question is how many fewer thunderstorms were available during the oily decade of the 80's compared to 50's & 60's certainly 2008 would stand out as a contributor to ozone sufficiency rather than a detractor. Please permit me a mildly penetrating question placed beside this factoid. "Ozone is created at the top of the atmosphere by the action of sunlight on oxygen". *How much extra ozone is created on the rising columns of Amazon rain forest oxygenated air? Or in a negative side to the question: How much has been lost by the drying and widening of the Sahara?* If we had real science in regard to this problem and real numbers as to the increasing incidence of skin cancer we could insert a few pounds of *elemental sodium or perhaps sodium hydroxide or even naked calcium,* at the correct altitude and observe closely for its effect as to the diminution of the proverbial ozone hole. It would seem to me quite natural if the offender on the threshold of space was a chlorine atom to send up his natural mate, sodium, and let nature take its course.

This observation is intensely relevant to the goal of carbon sequestration. You will recall an earlier mention under the category of Old Science (pre-1940) the 22 year solar cycle with its nadir marked by the hefty impingement of ultra violet light at sea level and its destructive effect upon ocean flora. The newer science notes only a reduced capacity for carbon sequestration at the South Pole because the flora suffers from ultraviolet exposure near the Antarctic icecap. Under either Science there is a deleterious effect.

It would seem to me that a fitting group of research questions could be orchestrated by asking for a comparison and contrast of peaks in the growth of CO_2 following hard on the heels of chlorine escape to the upper atmosphere. If CO_2 should spurt to new peaks two or three years following a spurt in chlorine releases it would seem to me a logical outcome in the topsy-turvy world of advanced biochemical toxicity.

Bikini Atoll Beginning Question Ocean Health & Atomic Fallout (To be revisited later)

That brings us in a roundabout way to the whole question of ocean health. This question could not have been raised in the tempestuous 60's. Sizably noisy and authoritative voices were almost insisting that nuclear fission? Fallout? energy? whatever was good for you. In 1968 an attempt was made to bring the Bikini Islanders back to their atoll. Bikini Atoll-Wikipedia, the free encyclopedia http://en.wikipedia.org/wiki/Bikini _Atoll.

"Bikini Atoll (also known as Pikinni atoll) is an uninhabited 6.0 square-kilometer atoll in one of the Micronesian Islands in the Pacific Ocean, part of Republic of the Marshall Islands. It consists of 36 islands surrounding a 594.2 square kilometer lagoon. As part of the Pacific proving grounds it was a site of more than 20 nuclear weapons tests between 1946 and 1958, including the first of a practical hydrogen bomb in 1952. Preceding the nuclear tests, the indigenous population was relocated to Rongerik Atoll. The tests began in July 1946. In the late 1960's and early 70's some of the original islanders returned from Kili Island but were removed because of the high radioactivity." (There follows a more complete and more ominous description of the same event.) "The Micronesian inhabitants, who numbered about 200 before the United States relocated them after World War II, ate fish, shellfish, bananas, and coconuts. A large majority of the Bikinians were moved to a single island named Kili as part of their temporary homestead, but remain until today and receive compensation from the United States for their survival.

In 1968 the United States declared Bikini habitable and started bringing a small group of Bikinians back to their homes in the early 1970's as a test. In 1978, however, the islanders were removed again when strontium-90 in their bodies reached dangerous levels after a French team of scientists did additional tests on the island. (1) It was not uncommon for women to fall ill and die during childbirth at a much higher rate than normal. The islanders sued the United States and were awarded $100 million in compensation.

The clean-up operation scraped off the top 16 inches of soil from the main island of Bikini, generating a million cubic feet of radioactive soil that could not be disposed of at a cost that far exceeds this compensation award.

Again from *http://www.bikiniatoll.com/whatrad.html* the IAEA's Bikini Advisory Group background radiation on Bikini:

"It is safe to walk on all of the islands . . . the Advisory Group reaffirmed: although the residual radioactivity on islands in Bikini Atoll is still higher than on other atolls in the Marshall Islands, it is not hazardous to health at the levels measured. Indeed, there are many places in the world where people have been living for generations with higher levels of radioactivity from natural sources such as geological surroundings and the sun than there is now on Bikini Atoll . . . But by all internationally agreed scientific and medical criteria the air, the land surface, the lagoon water and the drinking water are all safe. There is no radiological risk in visiting the lagoon or the islands. The nuclear weapons tests have left practically no cesium in marine life. The cesium deposited in the lagoon was dispersed in the ocean long ago.

The main radiation risk would be from the food, eating locally grown produce, such as fruit, could add significant radioactivity to the body. Eating coconuts or breadfruit from Bikini Island occasionally would be no cause for concern. But eating many over a long period of time without having taken remedial measures might result in radiation doses higher than internationally agreed safety levels.

Studies complete—time for recommendations

"Permanent resettlement of Bikini Island under the present radiological conditions without remedial measures is not recommended in view of the radiation doses that could potentially be received by inhabitants with a diet of entirely locally produced foodstuffs."

"While no definite recommendations are given on which strategy to follow, it is considered that the strategy using potassium fertilizer is the preferred approach. In this connection, it was noted that the soils of Bikini atoll are extremely deficient in potassium and extensive field trials have demonstrated that the application of potassium rapidly reduces the concentration of 137 Cs in food crops since potassium is taken up by the plants in preference to cesium. The reduction of 137 Cs in the food crops is sustained for about four or five years, after which the values slowly begin to increase again. However, repeated application of fertilizer forms an effective strategy in reducing the estimated doses to the potential inhabitants of Bikini Island. Furthermore, the supporting strategy of removing soil from dwelling areas would eliminate most of the external and internal exposures from direct soil ingestion or inhalation."

Well! What on earth does this material have to do with the health of the oceans? Not much except the $200 + million dollar price tag for 6 square kilometers of land in the mid-Pacific would when converted to an extended obligation for study and remediation of the world's oceans, translate to an economic cost exceeding the

gross national output of the five leading economic powers combined.

The unfinished question in my mind about Bikini Atoll is the lack of visible studies on the condition of the 594 square kilometer lagoon; we have only a rather casual reference to the increased fishing potential apparently caused by the shortage of fisher folk and an equally casual assertion that the radioactive cesium is by now thoroughly dissipated. But, *who can forget the televised footage of the (uniquely altered shark,) cruising the bottom of the bikini lagoon with a significant appendage missing apparently after a fallout altered genetic structure that bred true.* Where is Rachael Carson when we need her? DDT which has genuinely beneficial uses was banned on less evidence than we have before us about radioactivity.

If this nuclear material were on shore we might expect, hope for, and receive some reasonable and testable scholarship about the hazards; but we had to wait for President Clinton in the 90's to release 100,000 pages of incomplete data on the effects of nuclear fallout on American health. *But at this point I admit my negative bias in the case of nuclear fallout; particularly on and in the ocean, I feel a gigantic abyss of knowledge shortfall. This is admittedly one person's glimpse of the beast in the kitchen, but in regard to the health of the ocean it is still only one of several large and lurking candidates to be examined.*

Ice Age Triggers Nuclear? Volcanic? & Or Cosmic?

Since I am probably going to be perceived as anti-nuclear on the basis of the foregoing and subsequent material, I would like to take the opportunity to say some nice things about "Nuclear when viewed in a long term perspective". First, the obvious, we need the power. Second, we need the knowledge; the nuclear fires on the sun are not the biggest of our worries. We have nuclear fires in the earth that can be disastrously reawakened by long term magnetic changes of the earth, see Robert W.

Felix "Not by Fire but by Ice". The provocative truth that he drives home again and again in a powerful way is that Global Warming brings on the Ice Age, more accurately, OCEAN WARMING BRINGS ON THE NEXT ICE AGE. And the penetrating question which needs to be asked here is this: Is the Nuclear Genie safely contained in its magnetic bottle or is the decline of magnetic field strength and movement of the North Pole an indication of wonders and terrors to come that we can only guess at? Third, on a bit more ominous note; nuclear fission takes place in our upper atmosphere on a not so regular pattern whose dimensions are just now being fleshed out by cosmic ray observatories not the least of which, the Auger observatory in Argentina is just now coming to occupy a prominent sprawl in our index of fears during the years around 2012. In other words, we are all at risk of being baptized to a greater or lesser degree by the byproducts of nuclear fission. We might as well meet it head on with our best minds and our most cooperative research.

Magnetic Reversal Solar & Earthly? Volcanic?

If I understand Mr. Felix correctly the magnetic lines of force in the earth contain and separate the nuclear materials in the earth and serve to prevent run-away nuclear events in the same way that they serve to regulate combustion events on the surface of the sun. Then he notes the breakdown of these magnetic chambers on the sun and the resultant flares and/or coronal mass ejections; it is then not such a huge jump for him to extrapolate a similar result for earth. He speculates at some length as to the potential causes of magnetic reversals of which there are many proven episodes in our distant past. Recent movement of the North Pole 25 miles per year toward Siberia has been noted by the Kansas City Star. If this raises the spectre of the shadowy "beast" I would hasten to add that the North and South Pole of the Sun have swapped ends in the recent past (apparently every eleven years) with some but not massive scholarly linkage of the event to CME's or coronal mass ejections. Further if the banked nuclear fires are somehow unleashed by magnetic changes here on earth, we have yet to see a substantial number of Geiger counters being affixed to the 850 active surface volcanoes or to the process of assessing the growing census of undersea volcanic events. [Even the *scholars who dare to speculate on the causes of geothermal heat can't decide whether the leading heat candidate under the earth is radium or uranium.* At the very least we can appreciate that nuclear powered icebreakers may help us to cope with the sudden onset of a new glacial age.

> *In regard to "natural" as opposed to manmade radioactivity (Here dear Reader you must permit me a small possibly irrelevant question or two). If we were to study volcanic output both surface and subsurface for "natural" radioactivity, I would personally like to know whether there is a preliminary or an ending peak of radioactivity in the course of a volcanic event. Also, I would like to know where in the world the preponderance of nuclear emissions were? Further, if the studies had a sufficient duration in time whether radioactivity could serve as a predictor of imminent reoccurrence of volcanic activity. At the least that knowledge would aid in the harvestability of geothermal energy.*

The longer term benevolence of the earth's climate is intimately linked to our studied adaption to the almost unpredictable outbursts of volcanoes. A further reference to our perception of the invisible beast, I strongly doubt that you'll find it anywhere else. *If the Ring of Fire—the preponderance of earth's active volcanoes—is primarily a product of the immense gravitic and tidal forces at work over the huge span of the Pacific Ocean, then the melting of the Icecaps and the subsequent Ocean surge surely would lead to a predictable increase in Ring of Fire activity and thus to a heightening of global climate trauma.*

Before I do the unthinkable and switch sides (again) and spend considerable space introducing you to an energy pioneer and old family friend, perhaps the leading advocate of safe nuclear energy utilization, I would like to point out to you Sir Richard that you have the fundamental tools of carbon sequestration already in front of you. *What is missing is the chain of logic and ethics that will allow you to apply that logic in a careful and restrained manner so that we may back off from the tipping point of the climate that James Hansen writes so eloquently about. (Selections on "tipping point" will follow).*

Yellowstone & the Tipping Point! (First Visit)

One of the most dramatic candidates for the "tipping point" is the Yellowstone Park Caldera. It doesn't' quite qualify for the invisible "Elephant in the Kitchen" nomenclature because I have in my possession 93 pages of scientific data as well as some PURPLE PROSE about the forthcoming eruption. These selections are from the internet, *http://www.earthmountainview.com/yellowstone/yellowstone.htm*

"I have chosen parts which highlight the dilemma of the scientist who wears two hats; tour guide to this strange and wonder filled site on earth and prophet of doom. From page 3 to 93 '600 miles from the caldera is not safe at all.'

FEMA could not handle this big an event. The U.S. economy could come to a halt. Grocery stores would empty out, airlines, trains, buses, and roads would stop.

Though at the end of the film, during the last half hour Tom Brokaw and workers at Yellowstone Park tried to make it seem like nothing is going to happen, it is already known that the scientists who are monitoring what is going on at Yellowstone is (sic) being withheld from the public.

The real truth is, WILL NOT BE TOLD HOW BIG OR HOW BAD AN ERUPTION CAN BE, and the last word by the Yellowstone worker was, 'Come and visit Yellowstone and see how it really is.'

This is probably a good idea people "should" go see what is going on. Don't just go there and see how pretty all the steam and water is that is shooting up out of the ground. Find out the real science of the area and how it is changing.

Same source page 44 " Jackson, Wyo.—A network of sensors being installed to monitor earthquakes along the Teton Fault in Northwest Wyoming should be completed by the end of August 2002 or 2003 Seismologist Harley Benz said.

The seismic network of seven to eight monitors will help emergency response teams prepare for earthquakes, said Benz, who works for the U.S. Geological Survey at the National Earthquake Information Center in Golden, Colorado, the agency that is installing the network of sensors in Jackson Hole.

The new sensors will provide more data about earthquakes than a network of older monitors recently removed by the Bureau of Reclamation. Teuton County fought the BLM over removing the monitors but lost.

Benz said the new system will be able to tell scientists where the quake is, how big it is and the distribution of ground shaking." Back to page 3-"508 earthquake swarm in April, 2004." My comment—It's time to compile an average energy emission rpt. From pg. 73j of the same material. "Another potential hazard is from hydrothermal explosions that could send boiling water and rocks shooting into the air. The concern for public safety is real. Hydrothermal explosions have occurred recently at Norris and other areas of Yellowstone. For example, Pork chop Geyser exploded on September 5, 1989. Rocks surrounding the old geyser were upended by the force of the explosion, and some rocks were thrown more than 216 feet from the spouting geyser. Luckily, people in the area were not injured by the flying debris and scalding water. The cause of the increased thermal activity is not known, but scientists associated with the Yellowstone Volcano Observatory (YVO) launched a temporary monitoring experiment in August in order to learn from the ongoing activity. YVO is a collaborative partnership between the U.S. Geological Survey, the University of Utah, and Yellowstone National Park. The Norris monitoring experiment is also supported by two research organizations the Integrated Research Institutes in Seismology (IRIS) and University NAVTAR Consortium (NAVCO). Scientists installed a network of seven new seismic stations for recording various types of earthquakes. The instruments, called broadband seismometers, record a wide range of vibrations typical of hydrothermal and volcanic systems. These seismometers are especially sensitive to the long-wavelength ground vibrations that occur as water and gas move through underground cracks. Five high precision Global Positioning System receivers also were installed at Norris in order to track movement of the ground in response to underground pulses of groundwater and steam, and in case one occurs, a hydrothermal explosion. Data from the broadband and GPS receivers are being stored on site. The instruments and data will be retrieved in the next few weeks before the onset of winter. Thermometers were also placed in hot springs and downstream from geysers and other thermal features to continuously measure temperature fluctuations that may occur. The Norris experiment is intended to document activity within the shallow hydrothermal system that may be causing changes at the surface of the Back Basin. In the coming month, scientists will be pouring over the mounds of data collected by the Norris experiment for possible clues to the renewed heating of Norris. THERE IS NO EVIDENCE; HOWEVER, THAT MAGMA BENEATH THE ENORMOUS YELLOWSTONE CALDERA IS DIRECTLY INVOLVED. Scientists have noted similar changes at Norris in the past, but the current activity is perhaps the best opportunity yet to quantitatively document and better understand hydrothermal disturbances and their possible causes at Yellowstone. This article was written by scientists at the U.S. Geological Survey's Hawaiian Observatory."

The next article on the same page is apparently a little later—observations using the newer sensors. "Mt. Sheridan has been rumbling (15+ micro-quakes) between 1:00 p.m. and (9/7/03 between 10:00 a.m. and 1:00 p.m. MT (9/7/03), which were felt at Norris Junction. There were some small quakes between Midnight and 6:00 a.m. (9/7/03) at Norris Junction. There was a whole string of micro-quakes (25 or more) at Madison River between 6:00 a.m. and now, which are continuing. There have been sporadic micro-quakes (32+) all day at Mammoth Hot Springs. Micro-quakes started around noon and have continued to the present at Mirror Lake Plateau. All in all, activity is picking up from a lull for about two weeks, before which a series of small and large quakes (including a 4.4) occurred. That quake promoted the web report.

Steam pressure is apparently (is this science?) building again and hydrothermal fluids and steam are working their way up through fractures and vents. I do not expect anything unusual or extreme to happen in the immediate future, but if the trend continues, and the numbers of earthquakes gradually increase with **time, more warnings from geologists will ensue.**

What you should be alert to is any report that mentions increasing geyser activity, with new fumaroles and steam vents appearing near or on the top of the rising dome. The dome has risen three feet in the past years, and MAGMA HAS RISEN TO WITHIN 3.7 KM. OF THE SURFACE BASED ON QUAKE DATA."

Harvestability of Geothermal (1st Hint)

I think it's time to insert my semi-scientific (or pseudo-scientific) opinion into the middle of this conflicting scientific opinion to wit the previous boldfaced claim that magma is not involved or that it is only 3.7 km from the surface. Either way it's close enough for me to raise the commercial and industrial question. Is it close enough to be worth harvesting? The people whom I trust to perform the evaluation of this complex project are the people who have years of harvesting geothermal to form Iceland's backbone of power. I would ask a team from Iceland to join Chevron in making a feasibility study of Yellowstone's potentially harvestable power.

I WOULD ALSO INCLUDE A SUITABLE EXPERT ON THE RADIOACTIVE and Hi-temp AGING OF CERTAIN METALS THAT WOULD BE USED IN THE PIPING OF SUPERHEATED STEAM!

The question of whether this could become the backbone to an insurance policy against a "natural" return to an Ice Age, could be tabled temporarily. It would be resurrected in detail when Congress would consider afresh the necessary expansion of the electrical grid. The question of water availability will be discussed at length in another section. Rest assured, it will not happen without adequate water.

The emergency nature of the question of adequate water has to do with another assessment of the problem: to wit. The trouble begins, not with the constantly burbling water, but when it stops.

Size of the Threat of Mt St Helens vs. Yellowstone

I would like to return to common sense assertions with a little help from the enclosed pictorial presentation of former Yellowstone eruptions compared to the mapped zone of volcanic debris see map below. I was living 1,700 miles from the scientifically designated Ash fall zone of *Mt. St. Helens A few days after Mt. St. Helens erupted in 1980 I started out of my house to climb into my shiny new (red) Toyota Tercel, I noticed a red orange film on the trunk and wiped it off before climbing in and making my way to work. After breathing outdoor air that day I had a sore and inflamed throat for a week. The enclosed sheet claims that St. Helens ash zone extended 30 km. Then what was that stuff that I wiped from my car and tried for a week to remove from my throat?*

I'm not trying to quarrel with science here. I'm simply trying to plead from the potential energy release of Volcanic Yellowstone whether next year or 2074 or 200 years from now, any energy harvested and utilized would ultimately detract from the civilization destroying power of an un-harvested Yellowstone!!

Mt. St. Helens Ash Fall vs. Yellowstone

A cautious power solution would be to harvest the heat from the rear (southwest corner) of the slowly moving energy seep.

A gutsy (foolhardy) approach might attempt to harvest the advancing zone (northeast corner). See map of plate motion Hot Spot Track. (below)

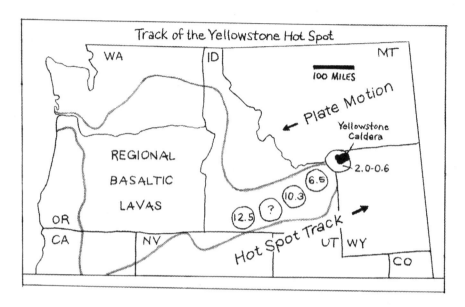

Track of the Yellowstone Hot Spot

But here I believe that I've already exceeded my area of competence. At the very least the chevron & Iceland energy experts could offer a limited feasibility study. Here we must suspend my musings and return to excerpts from YELLOWSTONE SUPER VOLCANO GETTING READY TO BLOW ITS CORK as previously mentioned. *http://www.earthmountainview.com/yellowstone/yellowstone.htm* pp. 73-74 "Earth loci measured to within 0.5 km under Mt. St. Helens, and people still didn't think it would erupt.

"But everything has to be scaled up for Yellowstone, meaning that 3.7 km is not a safe depth. Ground temperatures in the northwestern part of the park are apparently on the rise (up to 200 degree Fahrenheit in some places), killing the vegetation. Large areas of the park are now closed, including areas with geysers, because their water temperature is now scalding and dangerous for visitors.

"If more steam vents appear, that means a continuous pathway for pressure release has been established to the magma chamber. If that happens, the pressure in the magma chamber will continue to drop until it reaches a critical stage when the superheated water within the magma explodes. When that happens the super-volcano will blow violently, blowing out a chunk of its cap-rock and sending millions of cubic feet of ash into the atmosphere in a Pompeii-like explosion, but 100,000 times worse.

"When you hear those reports, you will have about two days to 'get out of Dodge' before the eruption. Unfortunately, as the **steam venting subsides there will be a false sense of security.** People will think it was just another cyclical event, and the danger is over. But that will be the farthest from the truth. *It will be the quiet before the storm.* A major earthquake will suddenly rock their towns for hundreds of kilometers around Yellowstone, and soon thereafter 1,000+ degree pyroclastic flows will descend on them at hundreds of mph, extending out to 600+ km.

"That 600 km radius around the caldera will experience total devastation. The next 600 km out may receive as much as 5-10 feet of ash, depending on wind direction. The thickness of ash will decrease away from the super volcano, but will reach the crop belt in the Midwest (Missouri, Iowa, Kansas, Nebraska, etc.) destroying most of the fertile croplands of the United States. California will be hard hit by falling ash, with its central wine valley severely damaged (the French will love it). Agriculture will have to shift east of the Mississippi for years. The Garden State will once again live up to its name.

"In northern Idaho you will have to contend with several feet of ash and isolation. Roads will be closed, power will be out, and phones will be out. Communication will depend on Ham radios and local stations that have generators. Rescue will take weeks or months. Some areas will never see rescue teams. The survivalists will be best prepared to make it through the difficult months following the eruption. Make new friends, have plenty of dust masks on hand, because you cannot breathe any airborne ash if you want to avoid lung disease. It's what caused mass kills of plains animals 12 million years ago, resulting in extensive bone beds beneath the ash. Drinkable water will sell at the price of gold. To recap, I don't expect anything to happen in the near future. But with such an unpredictable event, being prepared is your best ticket to survival."

Other Tipping Points, Meteors & Cosmic Rays

We have apparently backed into the important question of the "Tipping Point" while tentatively exploring the potential ramifications of "natural volcanism". So let's postpone our get-acquainted visit with Don Trauger the Atomic energy VIP, while we get acquainted with another VIP James Hansen who probably wouldn't protest too strongly about being associated with the term, only perhaps with the breadth of my application of its meaning.

One important non-candidate for invisible elephant of the "tipping point" will be dismissed with only this paragraph. Meteor or asteroid collision with earth has been amply dealt with in the media. With the world's best collection of nuclear missiles most reliable and accurate aiming devices and the world's largest (and growing) assortment of eyes on the sky, there is little I can add to this formidable arsenal of preparedness.

> *The other important celestial reason for delaying Don Trauger's introduction of the 'young lions at NASA', as the moving force for rewatering the Sahara. Only they have the science, and grasp of the 'heavenly' influence on water availability. Perhaps only a Nebraska farmer can grasp the delicious irony of multiple Sahara residents complaining about 'too much water! This next section calls attention to an area that they may hold almost exclusive competence.*

One little-known candidate for the tipping point is referred to in BBC news world edition Sat. October 19, 2002 from the internet—*http://news.bbc.co.uk/2hi/science/nature/2333133.stm*

"Cosmic rays 'linked to clouds' German scientists have found a significant piece of evidence linking cosmic rays to climate change. They have detected charged particle clusters in the lower atmosphere that were probably caused by the space radiation. They say the clusters can lead to the condensed nuclei which form into dense clouds. Clouds play a major, but as yet not fully understood, role in the dynamics of the climate, with some types acting to cool the planet and others warming it up. The amount of cosmic rays reaching Earth is largely controlled by the Sun, and many solar scientists believe the star's indirect influence on Earth's global climate has been underestimated. Some think a significant part of the global warming recorded in the 20th Century may in fact have its origin in changes in solar activity not just in the increase in fossil-fuel produced greenhouse gasses.

"First Evidence found: The German team, from the Max Planck Institute of Nuclear Physics in Heidelberg, used a large ion mass spectrometer mounted on an aircraft. They say their measurements 'have for the first time detected in the upper troposphere large positive ions with mass numbers up to 2,500'. They conclude: 'Our observations provide strong evidence for the ion-mediated formation and growth of aerosol particles in the upper troposphere.' The scientists report their findings in Geophysical Research Letters, a journal of the American Geophysical Union. They support the theory that cosmic rays can influence climate change and affect cloud albedo: the ability of clouds to reflect light.

"Cosmic rays and clouds: (The Sun's magnetic field and solar wind shield the Solar System from cosmic rays (very energetic particles and radiation from outer space). Changes in solar activity will affect the performance of the shield and how many cosmic rays get through to Earth. Theory suggests cosmic rays can 'seed' clouds. Some satellite data have shown a close match between the amount of cloud over Earth and the changing flux in cosmic rays reaching the planet.

"The importance of Clouds in the Climate system is described by the Tyndall Centre for Climate Change Research, at the UK's University of East Anglia (UEA). It says: "Clouds strongly influence the passage of radiation through the Earth's atmosphere.

'They reflect some incoming short-wave solar radiation back into space and absorb some outgoing long-wave terrestrial radiation: producing cooling and warming effects respectively.' And UEA's Climatic Research Unit spells out the complexity of Clouds' role in climate change. It says: 'The cloud feedback may be large, yet not even its sign is known. 'Low clouds tend to cool, high clouds tend to warm. High clouds tend to have lower albedo and reflect less sunlight back to space than low clouds. 'Clouds are generally good absorbers of infrared, but high clouds have cooler tops than low clouds, so they emit less infrared spacewards'. To further complicate matters cloud properties may change with a changing climate and human-made aerosols may confound the effect of greenhouse gas forcing on clouds. Depending on whether and how cloud cover changes, the cloud feedback could almost halve or almost double the warming.'

"Many scientists agree that the Earth's surface appears to be warming, while low atmosphere temperatures remain unchanged. (?)

Missing Link: Research published last August suggested the rays might cause changes in cloud cover which could explain the temperature conundrum. The discrepancy in temperatures has led some scientists to argue that the case of human-induced climate change is weak, because our influence should presumably show a uniform temperature rise from the surface up through the atmosphere".

"Although researchers have proposed that changes in cloud cover could help to explain the discrepancy, none had been able to account for the varying heat profiles. But the study suggested that cosmic rays, tiny charged particles which bombard all planets with varying frequency depending on solar wind intensity, could be the missing link."

*I have no quarrel with these insights on cosmic rays;
but I can't help but wonder when these scientists*

began their cloud researches. Different temperatures in different layers of the atmosphere have been around for a long time-some are no doubt decisively related to the massive clouds of dust and man-made radiation that were injected into various layers of the atmosphere depending upon the amount of energy propelling them into the more direct sunlight of the higher atmosphere. I suspect that volcanic clouds have a substantially different temperature but I could understand their reluctance to measure the temperature of volcanic grit in an airplane which could incur substantial harm from the glassy grit.

Brownian Movement Revisited
(North Pole Has More than South)

I can't refrain from sharing from high school physics re: Brownian movement. Dictionary definition Webster's 3rd new International dictionary Copyright 1966 (unabridged) BROWNIAN MOVEMENT OR BROWNIAN MOTION [after Robert Brown 1858 Scot. Botanist who discovered it]: the peculiar random movement exhibited by microscopic particles of both organic and inorganic substances when suspended in liquids or gases that is CAUSED BY THE IMPACT OF THE MOLECULES OF FLUID SURROUNDING THE PARTICLE. Perhaps it is just my slowness but the fascinating sight of the Brownian Movement in H.S. labs didn't begin to compare with the revelation that greeted me 12 years later while sweeping a dusty trailer at Yellow. There I noticed an immense difference in the time aloft of dust when it is impaled by a shaft of sunlight. The energy converted to heat by the opacity of dust was the motivation for the extra time aloft . . . and as a consequence the "peculiar random motion" is not so peculiar.

This same thought applied to the North Pole stratosphere gives a heartening warming not exclusively attributed to "Long Waves." That is then the proper foundation for the "Cure" to be anticipated for the South Pole Ozone Hole, but it does not cover all the bases of whose welfare will be affected both positively and negatively by the "cure".

Cosmic Rays Might Be One Trigger For Bringing Rain Back To The Sahara

Global Warming

"The epidemic that we face as a society today is not another nuclear holocaust, but that of global warming." We're presently in the middle of some nuclear effects related to cosmic rays. Let's highlight the size of the problem before we suggest some potential urgencies of a timely nature.

From Wikipedia—http://en.wilipedia.org/Greisen-Zatsepin-Kuzmin-limit (called ultra-high energy cosmic rays). "The observed existence of these particles is the so-called GZK paradox or cosmic ray paradox".

"Unsolved problems in physics: Why is it that some cosmic rays appear to possess energies that are theoretically too high, given there are no possible near-Earth sources, and that rays from distant sources should have scattered off the cosmic microwave background radiation"?

The question of cosmic ray distribution through time is uppermost in my own mind for reasons that should soon become apparent. Here again is material from the same Wikipedia source: "In 1934 Bruno Rossi reported an observation of near-simultaneous discharges of two

Geiger counters widely separated in a horizontal plane during a test of equipment he was using in a measurement of the so-called east-west effect. In his report on the experiment, Rossi wrote "it seems that once in a while the recording equipment is struck by very extensive showers of particles, which causes coincidences between the counters, even placed at large distances from one another. Unfortunately, I did not have time to study this phenomenon more closely." In 1937 Pierre Auger unaware of Rossi's earlier report, detected the same phenomenon and investigated it in detail. He concluded that extensive particle showers are generated by high-energy primary cosmic ray particles that interact with air nuclei high in the atmosphere, initiating a cascade of secondary interactions that ultimately yield a shower of electrons, photons, and muons that reach ground level."

"A huge air shower experiment called the Auger Project is currently operated at a site on the Pampas of Argentina by an international consortium of physicists. Their aim is to explore the properties and arrival directions of the very highest energy primary cosmic rays. The results are expected to have important implications for particle physics and cosmology."

Chapter 5

A CERTAIN DELICATE SENSE OF TIMING

The question of cosmic ray distribution and frequency as well as the looming problem of size should arrive center stage in the next few years with arrival of the long anticipated (dreaded? Feared? Keenly desired?) Solar System Crossing of the Galactic Plane in 2012. Why an event of this magnitude should have so precise a date certain is not clear but it is clearly laid out in at least two calendars, the Mayan and the (Hebrew?) **If I with my homespun lack of astronomical knowledge were to attempt to immortalize myself in my homebuilt flying saucer by crossing the Galactic Plane, I would allow at least a year for the event and maybe even a light year.** *But if I were to attempt to identify the critical time in that span, I would isolate that point in time that the sun, moon and venus conspired to bring earth to its closest approach to the sun!*

We can have a minimal certainty that none of us living has any prior experience of this nature since the last one was just a bit short of 26,000 years ago.

The reason for raising this problem at some length is that rainwater and sunshine are both indispensable components of any successful carbon sequestration event. It simply cannot occur without an abundance of both.

If we are to have a problem with cosmic rays in the very near future it may be in the form of a sudden **abundance of clouds with no predictable solution in** onshore useable rainfall for the selected—soon to be tillable and newly lush vegetation —whether it be Sahara, Mohave, Sahel, Arabian, steppes of Russia, and/or even the Gobi desert.

I haven't yet successfully linked (in your mind)—cosmic rays, Galactic Plane, and tipping point of the climate toward??—Let me do it first with just a hint of a suspicion that the Galactic Plane may provide us with an unexpectedly huge increase of cosmic rays emanating from the Black Hole at the center of the Galaxy. *And what kind of awakening will this bring to our own unusually quiet sun?* How much forewarning the Auger observatory and other cosmic ray observatories can provide is still an open question. And of course cascades of cosmic rays are linked to cloud formation.

Global Perspective on CO2 As An Asset! (Fact)

Since I am implicitly revealing my bias I might as well speak out directly. I am interested in sequestering carbon onto artful and useful as well as productive growth on the surface of the seven continents, particularly on those areas of the globe which have been short-changed by the processes of "normal" rainfall. In order to achieve this objective it will be acceptable to rejuvenate those sections of the ocean which for one reason or another do not presently surrender their moisture.

That this is a difficult or near impossible goal is revealed by this quote from the internet: *http://www.futurepundit.com/archives/002056.html* "The Ocean contains an estimated 40,000 billion tons of carbon, as compared to 750 billion tons in the atmosphere and about 2,200 billion tons on land. This means that, were we to take all the atmospheric CO2 and put it in the deep ocean, the concentration of CO2 in the ocean would change by less than 2%".

To put this in another light-if the world's deserts were somehow supplied with water-there is more than enough vacant—relatively speaking—(useless) land to fix all 750 billion of the airborne CO2 into the world's deserts with scarcely enough left over to supply the agricultural needs of the world's breadbasket located in the heartland of America. With that overwhelming super-abundance of raw material at our fingertips it would seem hardly necessary to build a Super-State with overwhelming restrictive powers. (WHOOPS! I SEEM TO HAVE SLIPPED INTO PURPLE PROSE-IT IS ALMOST NATURAL DON'T YOU THINK)! But seriously don't you think that a systematic attack upon the world's deserts is worthy of a little PURPLE!!!

Let's attempt to put a decisively different spin on this whole notion of CO2 as a toxin! I am quoting from ENVIRONMENT AND CLIMATE NEWS, AUGUST, 2009 p. 18 "Correction The July issue contained an error . . . [correct text follows] Some day the world will wake up and laugh when we finally realize that the entire pursuit of economic ruin in the name of saving the planet from increasing carbon dioxide has in fact been a terrible joke.

It is an unarguable fact that the portion of the Earth's greenhouse gas envelope contributed by mankind is negligible, barely one-tenth of 1% of the total. CO2 is no more than 4% of the greenhouse gas envelope (water vapor is more than 90%, followed by METHANE and nitrous and sulfur oxides). Of that 4%, mankind contributes a little more than 3%, divided among transportation, energy production, etc. 3% of 4% is 0.12. And for that we are sentencing the planet to decades of economic devastation.

Ironex as graveyard supervisor (or banking CEO) for Siberian methane.

Since methane is reputed to be between 23 and 100 times more influential than CO2, all that is required is that Mr Gore change focus a tiny bit and finance one or two timely trips of the ironex ship to the mid-Atlantic and seed the Gulf Stream just in time to enhance the spring and autumn precipitation for Norway and points east. We have already been treated to the sight of numerous Siberian lakes bubbling raw methane through the long Siberian winter. The proper interment (read banking) process for that Siberian methane is contained in that one ironex ship. (to be discussed more fully later.) The necessary carrot for the political process is to note that Sprengel's law when applied to northern Siberia can produce a monumental and growing bank of naturally sequestered and recycled CO2. The natural gas that accumulates beneath this interment process, when harvested, can only enhance Russia's position among the family of nations. This process is so simple and cheap with such tangible rewards for the future that it may swamp the reasoning process that brought this manuscript this far. Please allow me a chapter to properly complicate the process. It will appear to be a digression from cosmic rays except the discussion of cosmic rays is meant to underscore the need for the technical competence of NASA scientists. This following discussion will challenge not only NASA scientists but will need the enthusiastic cooperation of Russian scientists.

A Friendly Revisit To Radioactivity

Let's return to the subject of radioactivity in a friendlier manner. The internet http://muller.lbl.gov/teaching/ physics10/old%20physics%2010/chapters%20 (old)/-Radi . . . 11/20/2009 Introductory anecdotes no. 3 "The United States Bureau of Alcohol, Tobacco, and Firearms tests wine, gin, whiskey, and vodka for radioactivity. If the product does not have sufficient radioactivity, it may not be legally sold in the United States. No.2 "If you aren't radioactive, then you are dead. And have been so, for a long time. " If C 14 is the dead radioactivity, the time span is probably greater than 57,300 years. "Digression: Why we drink only radioactive wine and liquor. In the United States, alcohol for consumption must be made from fruits, grain, or other plants. It is illegal to make it from petroleum. (I don't know why, but it is the law.) Of course, any alcohol made by fermentation of plant matter contains recent, radioactive carbon-14. In contrast, petroleum was created by decaying vegetable matter that GOT BURIED 300,000,000 YEARS AGO. There is no detectable C-14 left in it. This difference provides an easy way for the U.S. government to test to see if alcohol was produced from petroleum. The United States Bureau of Alcohol, Tobacco, and Firearms tests alcoholic beverages for C-14. If the expected level of radioactivity from c-14 is present, then the beverage is fit for human consumption. If it is not radioactive, then it is unfit." There are two important reasons for examining the melting methane in Siberia for radioactivity and one is not to consider the option of converting the rotting vegetation into vodka. On the surface I believe that we need to know the age of the c-14 bubbling up through the Siberian winter, As a footnote to that thought, I don't quite believe that 300 Million year designation for the preparation of fossil fuels. It certainly doesn't qualify for a successful long term investment. The second question is even more important! Is there any indication that other kinds of radiation are producing an unwelcome surge of heat from beneath the frozen tundra? This question rises to dominant importance with the approach of 2012 and related years. Ever since I viewed the under Sahara pictures of long lost rivers and heard about the under ice topography of Antarctica and the deeper than 200 meters heating of the ocean, I have wondered how deep the much more powerful rays of the sun might exercise their influence, compared to the somewhat feebler but selective scientific instruments. I am nevertheless prepared to expand my admiration for the nuclear testing community, if they are able to give us straight talk about the kinds of radioactivity percolating up in the Siberian night. The rewards of harvestable methane with only a few decades of incubation time are just short of incalculable. As attractive as the North's options might appear, I remain committed to my first choice—to make the desert to blossom "like a rose".

Since my grandfather's timing of the purchase of Dust Bowl land was crucial toward the success of my Father's conservation theories it is necessary for us to investigate somewhat further into the question of timing both solar and cosmic as well as seasonal. From *http://en.wikipedia.org/wiki/Solar_cosmic_ray*

Concluding Cosmic Ray Comments

"Solar cosmic rays are particle radiation that originates from the Sun. Most are made of protons with relatively low in energy (10-100—keV; 1.6-16 fj/particle). The average composition is similar to that of the Sun itself.

The name solar cosmic ray itself is a misnomer since cosmic implies that the rays are from the cosmos and not the solar system, but it has stuck. The misnomer arose because there is continuity in the energy spectra, i.e. the flux of particles as a function of their energy, because the low energy solar cosmic rays fade more or less smoothly into the galactic ones as one looks at higher and higher energies. Until the mid 1960's the energy distributions were generally averaged over long time intervals which also obscured the difference. Later, it was found that the solar cosmic rays vary widely in their intensity and spectrum, increasing in strength after some solar events such as solar flares. Further, an increase in the intensity of solar cosmic rays is followed by a decrease in all other cosmic rays, called the Forbush decrease after their discoverer, the physicist Scott Forbush. These decreases are due to the solar wind with its entrained magnetic field sweeping some of the galactic cosmic rays outwards, away from the Sun and Earth. The overall or average rate of Forbush decreases then to follow the 11 year sunspot cycle. But individual events are tied to events on the Sun, as explained above.

"There are further differences between the solar and cosmic particles, mainly in that the galactic ones show an enhancement of heavy elements such as Calcium, Iron and Gallium, as well as of cosmically rare light elements such as Lithium and Beryllium. The latter are assumed to result from the spallation (fragmentation) of heavy nuclei due to collisions in transit from the distant sources to the solar system."

If that material is not confusing enough-there is a strong rumor among students of cosmic rays that there are high strength cosmic rays—possibly from deep space—colliding galaxies or even from the center of the Milky Way-that are capable of ignoring both the Solar Shield as well as the

Earth's magnetic shield. It's possible that we now have enough cosmic ray observations to make us aware of statistically significant bursts of cosmic rays but to coordinate this into a grand strategy for making "the desert to blossom like a rose . . . ." We note from the internet:

> *"http://en.wikipedia.org/wiki/Cosmic _ray.com* P. 3 Cosmic rays have been experimentally determined to be a potential modulating factor in cloud formation and by theoretical extrapolation to be a contributor of global warming. (3) It has been shown that cosmic rays have a catalytic effect on the nucleation of cloud droplets (3). It is similar in concept to the operating principles of the Wilson cloud chamber and the cosmic rays catalyze the production of Cloud condensation nuclei". But we're not quite ready to propose a program.

> *You need to first be face to face with my bias; which is, I favor almost any onshore rainfall short of a full-fledged hurricane. The reviving of 'normal' oceanic rainfall on land when coupled with its proper welcome on land is the aim. The fact of warmer than average oceans coupled with cosmic rain enhancers makes the choice of timing for rain enhancement a galactic event lest a careless scheduling might produce a public tipping point—against rain and for 'deserts.'*

Next we turn to James Hansen and his tipping point. Actually the popularity of the phrase is his but this work belongs to Mark Lynas.
http://www.Carboneequity.info/docs/3degrees.html.com—selections will be in quotes with my comments interjected.

James Hansen's 3°Tipping Point

"Drawing from his work and other sources, the following is an overview of some of the scientific projections as the world warms. Three degrees may be the "Tipping Point" where global warming could run out of control, leaving us powerless to intervene as planetary temperatures soar. America's most eminent climate scientist, James Hansen says warming has brought us to the 'precipice' of a great 'tipping point'. If we go over the edge, it will be a transition to 'a different planet', an environment far outside the range that has been experienced by humanity. There will be no return within the lifetime of any generation that can be imagined, and the trip will exterminate a large fraction of species on the planet."

I realize there is a great temptation to exaggerate for effect in a world supercharged with media and CO_2 laden exuberance and subsequent adrenal exhaustion but **The** tipping point toward the next ice age is the one I'm worried about. See previous "Yellowstone About to Blow." From the timeline of environmental events *http://en.wikipedia.org/?????* events—19th century—1815 Eruption of Mt. Tambora in what is now Indonesia, largest of Holocene. Followed directly by 1816-"The year without a summer." So it should be clear that almost any super-volcano or Nuclear War for that matter can send us a long way toward the next Ice Age—Then the matter of grave concern is that which will drive us the other way toward desertification, ice cap melting, oceans rising and extremely powerful hurricanes.

LOOKING FOR THE BIG PICTURE ON HURRICANES.

Let's deal with the last one first and see if it yields a bit to the application of common sense. Did you know that there were zero (until 2004 or 2005) hurricanes in the South Atlantic? I asked my Internet expert to look one up for me and after extensive searching; he may have come up with one 5 years ago. He did provide me with maps for the

Hurricane tracks in the North Atlantic for the last three years-quite a contrast with the emptiness of the South Atlantic.

I suspect that it would be a good idea to call attention to the real cause of that welcome absence of hurricane destruction— in addition to the obvious—sea temperature never got above 74 degrees till just a few years ago. It's listed at 80 degrees now which is indirectly traceable *(by me anyhow)* to the Ozone Hole usually ascribed to Antarctica. The not so obvious cause is that there once was a natural sweep of plentiful quantities of cold dry air off of the Antarctic Ice Cap moving up the South Atlantic Basin providing the impetus to sweep evaporated moisture off of the Atlantic onto first the African shore—supplying generously the Congo Rain Forest but continuing across to the mountains of Ethiopia where it generously contributes to the headwaters of the Nile Rivers. The Natural push of moist air westward toward the slopes of the Andes calls attention to the most August and prolific of all the weather systems of the earth resulting in the mighty Amazon whose vegetation *colored water proceeds from* its 150 mile wide exit into the Atlantic Ocean and continues to widen after the mouth and carrying fresh water almost to the coast of Africa. Such is its bounty that semi-serious proposals have been made to trap the water and pump it wholesale onto the Sahara in one more ambitious reclamation project. Having come this far, let me say aloud, the obviously logical next step. **"Let's get rid of North Atlantic Hurricanes the same way!"** *You'll notice that I didn't' say that too loudly! For one thing it runs counter to the present melting condition of the Arctic Ice Cap. But also it runs counter to the long term projection for earth's orbit which in short calls for warmer Northern Hemisphere winters and cooler summers. Fortunately for my next statement—authorities are divided as to whether the Milankovitch cycles require North Pole icing. I would bet a substantial amount of money that a well paid government climatologist would reverse the cause and effect portion of that sentence. But give me a little rope and I'll start to build a noose and maybe one big enough to hang Robert W. Felix along with me.*

This idea also runs counter to the controversial and since aborted cloud-seeding experiments of the 60's. The idea there was to seed a certain portion of the cloud and hope to alter its direction and hopefully minimize its damage. Needless to say that idea did not start soon enough in the temperature enhancement process-once the energy is there and activated no power on earth is going to form and shape and direct it to its maximally productive destination. I intend this manuscript to make public the means to harness that power and lessen its damaging consequences.

Even if you jump to the obviously preordained conclusion dictated by the anecdotal account of Antarctic Polar Air you will still not be early enough in the process to adequately comprehend Cause and Effect. Let's take an imaginary sequence in which there is a plentiful supply of ice & icy air on the North Pole. What on earth would dislodge it from its resting place atop the sea level Arctic Ocean? And in addition, which route would it take in its return to the equator? I'll give you a small logical hint, the Alps in southern Europe would at least be a small obstruction. There are two other choices (probably more but this is a caricature so let's keep it simple.) Besides the Atlantic Ocean there is Siberia and the cold records set there over the years indicate that this is indeed a prime path; and the leafless and mostly treeless center of North America—N. Dakota, S. Dakota, Nebraska and Kansas etc. Probably by now the obvious choice would be for the cold refreshment to remain at sea level cruising down the Atlantic Ocean.

But I've already told you that we didn't start early enough in the causal sequence, so let me blunder on weaving the rope that you may use to hang me. Where? (And When) does the air come from that thrusts the super chilled mass of air from its stationary site on the North Pole? We might have a hint in the relatively late arrival of the term Hadley cells, that attempts to describe the descending interruptions of heated equatorial air as it courses downhill from the solar inflated heights at the equator to its low lying accumulation at the North Pole.

EQUATORIAL SOURCES OF THE POLAR AIR MASS AND THE QUESTION OF THE EARLIEST TIMING.

Since we know that the air is inflated at the equator and then courses generally downward to the poles; the next question to be asked is "What inflates it to give proper impetus toward the poles?" The answer of course is the Conversion of sunshine to heat which occurs in various ways along the equator. We are going to temporarily ignore the most important of these El Nino in the Western Pacific—because at this point it will appear to be out of our control.

Still Looking For the Big Picture On Hurricanes

We are going to wrestle with four factors which we may be able to exert some catalytic control over with positive results. Number one is the high degree of reflectivity of the Saharan Desert. I list that as a negative-for now-. If you recognize the recent anemia of the Polar Air Flow south along the Atlantic as a contributing cause to the more frequent and devastating hurricanes then your logic is progressing nicely with mine.

The second factor is greening of the Atlantic Ocean outside of the Amazon River and continuing to its logical extreme in the Sargasso Sea-East of the Bahamas-size (from year to year) uncertain-transpiration rate compared with breezier portions of the ocean (unmeasured to my limited knowledge). But at some point its role as a semi-stationary heat sink should receive an evaluation for its role in relation to hurricane genesis and augmentation. Because this greening increases the rate of solar conversion, transpiration and subsequent evaporation I rate it a net positive. Threats to the Rain Forest of Brazil are seldom extended this far into the Ocean for factor 3—at least in published material, but the death of the Rain Forest would have more serious consequences than just reducing the world's supply of renewed oxygen. When the concern about greening of the Ocean is extended further down the Atlantic to include the ultraviolet disruption of the food chain around the Antarctic we almost have a handle on the size of the problem.

Even the Mostly Welcome Work Of The Corp OF Engineers Comes In For Mild Scrutiny

The threat of global warming gets closer to genuine fear here than any other point that I can think of. It deserves a further look under the 4th factor which is labeled the Gulf Stream and we need a bit of continuity. But instead of fear I see opportunity. Some authors suggest that the Gulf Stream has its beginning in the Pacific Ocean and progresses west from Australia around Africa at a speed of 5 mph. proceeding up the Atlantic to fulfill its destiny in the highest evaporative zone on the globe. The halfway point of its lengthy journey is the Northwest coast of Africa which is often noted as the beginning of Atlantic hurricanes. It proceeds westward along the northern coast of S. America where in 1922 it picked up some of the 900,000 barrels of spilled oil from Lake Maracaibo, Venezuela. From there it wraps around the Gulf of Mexico where it adds the floating contents of the 2nd longest contributor to the Atlantic. What that material is and how its contents have changed from the 20's and 30's to its present content and influence for today should be the object of at least as extensive an historical scrutiny as has been offered through the internet about the Mediterranean Basin. I don't presently have access there even though I possess an extensive history of the achievements of the Corps of Engineers. **an evaluative semi critical note should be inserted here.** The early history of the Corps of Engineers is laced with water entrapment schemes but has subsequently evolved into a road building operation and other structures (which may make it a piece of the problem. This relates to excess surface hardening). Back into the Atlantic and diverting north toward Labrador and from there-wind-aided across the Atlantic to bring warmth and moisture to England and spurring trees to grow out of nearly vertical rock formations in Norway. I ask the commonsense question-Why does this current stay on the surface for so long when a similar heat applied in the Mediterranean results in a subterranean exit from the Med of the saline water? I submit that the organic matter contained in the current had a somewhat significant relationship to this lengthy process. The health of the Atlantic conveyor system is affected in several crucial ways, not only by the poisons & toxins of the Mississippi but by the presence or absence of organic green flowing from the Everglades.

The logical speculation that could follow this description of a constipated Atlantic evapo-transpiration cycle is this: Whether caused by temporary oil sealing of the ocean or resulting soil hardening on land, the end result could be a prolonged hesitation of the cold dry air atop Antarctica, that once used to fall off of the icecap in the direction of the Amazon, the Congo, and the Niger. There are of course other factors which attempt to explain the 6+ degree temperature surge in the South Atlantic but I like this one because it becomes the capstone which reinforces the multiple efforts to find a cure for the 50 year + widening of the Sahara Desert.

Mediterranean Basin As A Prime Symptom Of Water Distress

So far we have what my son-in-law would label with his helpful get-to-the-point briskness is a succession of rabbit trails. For those who have never hunted rabbits a day or so after the snow-it's almost impossible to know which way to go. But for me this next factor is the grand unifying theme: however, unless you appropriate my father's summary of 30's weather:—"It's not as if we had a striking amount of rain difference in the 30's. It is that the moisture that fell-struck sun baked earth and ran away before it could soak in:." You'll get lost in the late cyclones of data from the Med.

> *You have my permission to skip the next few pages with just this brief explanation. The new Med is afflicted with erratic excess storminess and periods of extended drought.*

When you apply that same logic to the Mediterranean Basin, you have to amplify it by the presence of 274 million people and all of the hardening appurtenances of modern civilization-streets, highways, roofs, parking lots etc. By now you have access to most of the fundamentals of climate change. The following selections from HyMEX) "Hydrological cycle in the Mediterranean Experiment) is a major experimental program aiming at a better quantification and understanding of the hydrological cycle and related processes in the Mediterranean, with emphases put on high-impact weather events and regional impacts of the global change including those on ecosystems and the human activities. Http://www.cnrm.meteo.fr/hymex/

HyMEX aims at producing a **new long-term and highly temporally and spatially resolved data-set** over the Mediterranean basin to:

1. Provide an accurate description of the water cycle and its variability and trend (accurate documentation of different terms of the water budget over the different compartments and at their interfaces; documentation of the key process driving the water cycle.)

2. Understand how the Mediterranean water cycle processes contribute to the regional climate (explore and model the various mechanisms determining the space and time variability of water budget of the Mediterranean region; relate the regional

mechanisms to the large-scale circulation systems in the atmosphere and oceans over the globe.

3. Validate the regional oceanic, atmospheric and hydrological models and develop improved parameterizations. HyMEX also aims at developing methodologies and models in order to contribute to basic needs of weather prediction, regional climate studies, climate impact, and environmental research by:

1. highways, roofs, parking lots etc. By now you have access to Determining and/or improving the predictability of the water cycle, its variability and associated high-impact weather events.

2. Performing regional climate change scenario HyMEX focuses on the interactions and feedbacks between the various compartments (atmosphere, sea, continental surface and interfaces) and thus associates major disciplines such as meteorology, oceanography, hydrology and climatology. In particular, HyMEX addresses key issues related to:

   a. the water budget of the Mediterranean Basin
   b. the continental hydrological cycle and related water
   c. resources heavy precipitation and flash flooding
   d. intense air-sea exchanges
   e. coastal dynamics

Water budget of the Mediterranean basin: The Mediterranean Sea is characterized by a negative water budget (excess evaporation over freshwater input) balanced by a two-layer exchange at the Strait of Gibraltar composed of a warm and fresh upper water inflow from the Atlantic superimposed to a cooler and saltier Mediterranean outflow. Light and fresh Atlantic water (AW) is transformed into denser waters through interactions with the atmosphere that renew the Mediterranean waters at intermediate and deep levels, and generate the thermohaline circulation. Although the scheme of this thermohaline circulation is reasonably well drawn, **little is known about its variability at seasonal and**

interannual scales. For example, a better understanding of the formation of Levantine Intermediate Water (LIW) in the eastern Mediterranean is needed because LIW plays a major role in the formation of other dense waters in the whole Mediterranean (its signature is still visible in the Mediterranean outflow at the Strait of Gibraltar). Also visible are the feedbacks of the Mediterranean basin on the atmosphere through the terms of the water budget. The budget of the Mediterranean Sea has also to be examined in the context of the global warming, and in particular **by highlighting the impact of an increase of the Sea Surface Temperature (SST) on high-impact water frequency and intensity.**

Hydrological continental cycle: The rainfall climatology of the Mediterranean region is characterized by dry summers frequently associated with **very long drought periods, followed by fall and winter precipitation that are mostly very intense. This results in high daily/ seasonal variability in aquifer recharge, river discharge, soil water content and vegetation characteristics—for which the interaction with the atmosphere is not well known.** This included for example the impact of the large extension forest-fires associated with drought during summer on the evapo-transpiration component of the hydrological cycle. The role of the surface states (land use/land cover) and of the soils on the modulation of the rainfall needs also to be better understood. Hydrological and hydro geological transfer functions are also characteristic of the Mediterranean basin, notably because of their specificities of the peri-mediterranean karstic and sedimentary aquifers. Progress in their understanding is of a primary importance for the development of integrated management of the hydrosystems, and its adaption to anthropogenic pressure and to the climate change.

Heavy precipitation, flash-flooding and flooding: During the fall season, the western Mediterranean is prone to heavy precipitation and devastating flash-flooding and floods. Daily precipitation above 200 mm are not rare during this season, reaching in some cases, values as exceptional as 700 mm recorded in September 2002 during the Gard (France) catastrophe. Large amounts of precipitation can accumulate over several day-long periods when frontal disturbances are slowed down and strengthened by the relief (e.g. Massif Center and the Alps), but also, huge rainfall totals can be recorded in less than a day when a mesoscale convective system (MCS) stays over the same area for several hours. Whereas large scale environment propitious to heavy precipitation is relatively well known, progress has to be made on the understanding of the mechanisms that govern the precise location of the anchoring region of the system as well as of those that produce in some cases (an) uncommon amount of precipitation. The contrasted topography, the complexity of the continental surfaces in terms of geology and land use, the difficulty to characterize the initial moisture state of the watersheds makes the hydrological impact of such extreme rainfall events very difficult to assess and predict."

Summary my version

I have a very simple answer to this long and involved careful statement of the problem: Like all simple answers it deserves to be attacked from every angle under the sun-and some from beyond the sun-beyond the solar system. It does not matter much to me whether "Ocean Sealing" terminology is applied to the problem or whether "toxic sludge from human wastes or agricultural fertilizers or pesticides", inhibit normal rainfall events. The *ancient culprit, now the somewhat anemic hero,* is the lack of a driving cold dry air mass to initiate the cycle and carry the evaporated air mass far enough from the Sea (early enough in the growing season to augment the solar conversion rate produced by dark wet-black-soil and growing—dark green-vegetation). Just one patch of dark green or black wet patch won't do it. We've already documented the skewing of the rainfall cycle toward the latter end of the growing season.

I'm getting perilously close to the "Spit it out or get off the pot!" portion of this section. I haven't involved

Robert W. Felix appropriately yet but I think it's better if I take the blame for this idea . . . . We refer back to the Where? And When? Of the super-cooled air mass. Origination only for now please! And of course it originates with a rising column of heated air far enough north of the equator to make its North Pole destination certain.

We've already named four factors but before I answer the question of the 'timefulness' of their influence, I would like to refer you to an old standard reference so that you may properly value the singular nature of the contribution or delay of the influence that may be needed for proper and timely rainfall on the Mediterranean littoral or for that matter the reduction of the September and later abundance of Hurricanes in the Caribbean and Gulf States.

Bypass of Upper Nile Swamps To Promote Lower Nile Productivity—retards the earliest seasonal transpiration

Encyclopedia Britannica (1965) vol. 16 p. 523-Nile River mid-20th century water reservoir improvement plans-"Third a reservoir in Lake Albert was required to control water from Semiliki River and the large quantity coming in seasons of unusually heavy rainfall from the tributaries of Lake Kyoga and the amount of water sent down to the Sudan and Egypt. Fourth, in view of the losses of water in the swamps of the Bahr el Jebel, as it was obviously useless to provide large storage reservoirs, if half their outflow would be lost, the Jonglei diversion canal was designated to bypass the swamps; this would leave the Bahr ez Zeraf and the Sobat. A regulator would divert about half the discharge down this canal and the remainder would flow down the Bahr el Jebel at a level that would reduce the losses to normal, so that there would be a gain of water in addition to the regulated distribution produced by the lake reservoirs. The Jebel Aulia reservoir was to continue to act as it already did and to store water mainly from the Sobat Flood.

I have chosen this example of humanly reduced transpiration to consider the problem of solar conversion and transpiration rates separate from the modern PC of Wetlands and of oil spill pungency. By the way an interesting fact-(survey) of the Nile was included in the 1976 Guinness book of World Records p. 135. **Subterranean River.** In August, 1958, a crypto-river was tracked by radio-isotopes flowing under the Nile, with a mean annual flow six times greater-560,000 million cubic meters (20 million million cubic feet). One more significant question! Did the building of the Aswan High Dam reduce the flow of this mighty crypto-river? The answer of course is yes. And the reduction is so severe that a semi-serious proposal is on record to dynamite the Aswan Dam (destroying the livelihood of 20 million Egyptians) or damaging the med at Gibraltar all to save us from the next Ice Age in (2090)?

Chapter 6

The "new" (or periodic recontinuation) ICE AGE—to be or not to be? That my friend is the Question!

Part of me yearns deeply to embark on some "PURPLE PROSE" and try to accomplish in two or three paragraphs what it took me 50 years of mulling over. Rest assured that if you are already there I may, be curious about the tortuous path by which you arrived there. For the skimmers among you I will attempt the short form and then try to expose a bit of the painful logic as it accumulated over the years.

The Atlantic's Connection To Continental Water Distress (Productive Hints)

The decisive point on the globe where the human influence on the New Ice Age will be won or lost is the North Atlantic Basin-specifically the Sun Bowl north of S. America and extending from the Eastern Mediterranean to the far West extent of the Gulf of Mexico. It is punctuated on both of its extreme ends by practically unlimited oil (read carbon) wealth but its main challenge is to supply evaporative H2O to the almost unlimited evaporative zones around it. The heart of this Zone is the Sargasso Sea within the Bermuda Triangle and the Mystery beneath it is the almost uncounted numbers of volcanoes in partly known stages of activity. (Increasing or decreasing)? This zone is more than ably supplemented by the **murky green (solar augmented) runoff of the mighty Amazon.** Its once able partner in supplying surface darkening vegetation (more solar converting material) the **Mississippi, Ohio, and Missouri triumvirate have been compromised not only by dams & levees but by undigested organic wastes complicated by pesticides, fungicides, and all of the toxic and semi-toxic debris flowing from the chemical horn of plenty headed by the medical and pharmaceutical branch, (not to mention the military-industrial complex) of our society. The pulse of this mighty solar engine has been weakened in recent years in tandem with the much discussed shortfall of Arctic ice. My contribution (besides dividing global warming into two parts-Ocean Sealing and the related increased radiative access through the crippled Ozone layer) is to push the complex of causes one** (or more) **step** (s) further back to note the decline in seasonal greenery **(solar conversion pulses that supply the initial energy at the equator) which creates that surplus of rising hot air that cools as it proceeds northward and eventually falls near the Pole displacing the gathered cold mass southward** (hopefully hugging the Atlantic Basin) **to fulfill its proper function-to displace Atlantic moisture & heat before it accumulates to destructive Hurricane** (Gulf) **or Cyclonic** (Mediterranean) **force.**

I realize that this is a grossly inadequate picture (caricature) of an enormously complicated weather picture. My apologies in advance to the entire group of doctorate qualified climatologists whose work I have seemingly threatened or short-circuited with a North Pole perception that ignores Canada, the Pacific, and all of Siberia. I only desire at this point to highlight potential areas where human choices could contribute to a catalytic renewal of the cold dry air pulse that is the heartbeat of successful Atlantic evapo-transpiration.

For those who feel (like me) that I have short-changed the intellect in this cartoon—like portrayal of complex forces: I'd like to suggest some wholly Atlantic conundrums that could use further illumination as to the important question "Why"? First, the three year succession of hurricane maps of 04 and 05 followed by the unpredicted paucity of 06 hurricanes. This is to be followed by the trebly important question of 07? First, the Georgia-Florida early season forest fires with their extensive ash fallout onto the (weakened!?) Gulf Stream, Second, the truly massive almost unprecedented non-hurricane caused flooding later in the season on the perennially water-short Zone of Texas, Oklahoma, Kansas, and Southwestern Missouri. Third, the question-did Norway or Tex-Oklahoma receive the rainfall that the Southeast United States felt was their due? These questions may subside to lesser importance if the events of this year contribute to a substantial renewal of the strength of the Gulf Stream.

Beginning dreams of a more productive continent—ocean interaction

Since any proper scientific study would take years and mountains of Federal dollars-I'll offer my Dad's common sense answer-build a deep grass reserve to slow the water and retain it as high up the slope as possible-also terraces, impoundments, ponds, lagoons, wetlands,—even mountain peaks will do in a pinch for water storage places-providing the protective snow cover is not artificially triggered to a lower more erosive and evaporative level by the ambitions of skiers. **WHOOPS! I SEEM TO HAVE STRAYED FROM STRICTLY Atlantic Questions!** Permit me an additional question. This one is mainly Atlantic in its reference but its overtones reach to the late season Southern California forest fires as well as to the unusual Pacific influences. If the answer to the ash fallout on the Gulf Stream was that it carried excessive evaporation potential to Norway-resulting in markedly large late summer precipitation—remember I was there for two weeks in the middle of August and it rained every day-The next question is what would have happened if the fires had reached their peak a month or two later? Would that have been the missing precipitation necessary to ignite a more normal Northern Winter? And do the extensive Southern California late season fires provide the beginning glimpse of a new cooler normal? We'll find out! (trip taken in August of 07.

It's now early Dec 09 in the middle of a blizzard and I think we just found out!

I may have just strayed from the semi-technical zone into the warmer political waters. I may not be able to escape unscalded. The question of how hot you want the climate to be is too large for my capacity. I only wish to encourage more adequate rainfall for the water-short semi-arid areas of the world. *By accepting that goal I may be demoting to second place a simple, cost effective and profit-making alternative which holds the exceedingly attractive option of almost unlimited carbon sequestration capacity.* In achieving *the first-named goal,* human caretakers-horticulturalists or settlers could take over the nursery job for a billion or more new trees as they presently do for 47,000,000 new trees in Nigeria and Kenya. In doing that it will allow aware "Rain-Harvesting" societies to achieve Sir Richard's announced goal of sequestering a billion tons of CO_2 per year-particularly since it could reignite a more timely succession of rainfall to enhance a more normal rainfall during the growing season for other parts of the planet. If there is a public consequence to the application of these ideas, it might appear in the media as a nearly equal balance of semi-hysterical fears-on the one hand sudden onset ice age-on the other rising oceans and expanding deserts. If a happy medium is to be found it will appear in those years that new records for highs and lows will not be set in the same month or other equally short span of time. If the growing public perception of the solar conversion potential of the murky top foot or so of the Atlantic Ocean turns into public hysteria over the declining organic green being emitted by the Amazon into the Atlantic; then the much-discussed iron filings greening agent or other acceptable micro-nutrient could be added in a timely mode and in judicious quantities to supplement and enhance the life-giving qualities of the mightiest river on earth. We will deal with iron filings as a greening agent in the context of the successful experiments done with iron filings and we will highlight the need for on shore cooperation to complete the success of the phytoplankton enhancement.

I am going to hesitate in this ongoing monologue in order to respond to some of the questions which have undoubtedly arisen in the reader's mind. The first question, how dare you assert baldly that one relatively tiny area of the globe holds the key to success or failure of the world's climate? I will answer in part that I lack documentary proof of my assertions at this point so I will share from my beliefs which are backed by the memorable certainty that the documents are out there somewhere and can be found by persons with more access to historical documents than I. The arrogance that you see may be because I appear to be neglecting what most climatologists regard as the dominant feature of global climate-El Nino or La Nina-either Southern Oscillation or Northern Oscillation. I'll have a little to say about it later that may help but for now let's call *it* the joker in the deck. *I will also have something useful to say in case my climate prescription, when followed, produces a huge decline in the destructiveness of El Nino. This advice is specially aimed at the Prospective "Big Loser".*

Agulhas Demystified. Water Runs Downhill As Does Icy Water and Air

The evidence that I am missing probably accompanied the historical summary of the work that produced the Panama Canal-completion date-1914. Beyond the verifiable assertions that tides on the east end of the Canal average 2 to 3 feet in height and that the tides on the Pacific side average 17 feet in height lies the apparently unverifiable assertion that "If the Panama Canal were a sea level canal-there would be a four foot waterfall from the Pacific to the Atlantic Ocean." There I have said it and God knows that I need help from you Internet scholars and Canal historians to supply some authority to this central assertion. This assertion does not stand alone. I began this composition with some confusion as to the origin of the Gulf Stream possibly believing that its origin had been traced back to near the North Coast of Africa-that may have only been traceable origination eddies of hurricanes. During my searches I encountered another assertion that there is a stream of 5 mph that carries Pacific water west from Australia to the south of the Cape of Good Hope into the South Atlantic. **The Agulhas Current.**

There are two possible forces to drive this awesome river. The first is the well-known circum-polar vortex-from there we can understand a mass of cold air slipping off of Antarctica and turning the current northward toward the equator. But that current is going in the wrong direction. **But I'd like to suggest an alternate power source, particularly for turning the corner and heading north up the south Atlantic. The only other power strong enough to produce that current is the well known fact that water tends to run downhill. My question at this point is who established this speed with how many years of studies and how great were the seasonal as well as year to year variations in speed as well as breadth and depth of current. In other words-the change of pace of the current flow would supply some indelible guidance to the year to year and season to season rate of change of the evaporation rate of the Atlantic Ocean-whose main Solar exposure is the previously mentioned Sun Bowl north of South America.**

Searching Past Ages For A Not-Too-Dramatic Norm. Buried Peat Bogs & Coal Swamps

Before we establish some artificial norm we need to bear in mind some other variable norms that have changed with the passage of **certain dates on the calendar. The Dead Sea**-from previous Ency material—evaporation rate listed at one point as a variable 10 to 15 feet—seasonal and at another point-in the same article as approximately 10 feet. Another source which I cannot put a name to right now lists the total year-long evaporation rate at 22 feet. Additionally-"From 1879 to 1929 however, the water level rose only 23 feet, (a lesser rate of rise from previous times) and since then has dropped a few feet. (1965 stats) There is no commonly accepted explanation for this phenomenon." **(If you accept my scenario of the Gulf Oil Glut having its approximate genesis** in 1922 and its subsequent distribution through the North Atlantic in the years immediately thereafter, you don't have to wear yourself out looking for a non-existent consensus). But we are here searching for the potential norm for the necessarily variable rate of evaporation of the Atlantic Basin.

Let's try a somewhat lengthier look for a norm from *http://en.wikipedia.org/wiki/Dead-Sea* pg. 2 of 6-"From 70,000 to 12,000 years ago the lake level was 100 to 250 m higher than its current level. This lake, called 'Lake Lisan', fluctuated dramatically; probably to levels even lower than today. During the last several thousand years the lake has fluctuated approximately 400 meters with some significant drops and rises."

If you are looking to establish too narrow a norm better think again. But the wetter Near East coincided-timing wise-with the last glacial epoch. From Ency Britannica vol. 10 page 440. "So far as can be determined by use of the radiocarbon methods for dating wood and other organic matter found in glacial deposits, the last major ice advance in North America and Europe culminated about 18,000 years ago." But most of us would agree that the tipping point for that period of time which included the extinction of the wooly mammoth lies outside of the norm which we would seek to establish however non-restrictive we desire to appear. The mystery of the Siberian Mammoth found quick frozen with buttercups in his stomach-apparently at the end of an Icy Period-I confess that I'm still lost in the varying theories of that death. *One of the relevant questions that remain undealt with is this? Do buttercups now grow and blossom at the same latitude where the frozen mammoth was uncovered? And did those buttercups prosper under the somewhat colder conditions prevailing in the decades following WW2?*

If you are still holding the imagery of the Antarctic air mass slipping slyly North at the wave tops of the Atlantic-driving moisture off of the Atlantic feeding both the Congo and Amazon Rivers and still having enough moxie to form the headwaters of the Nile River; then you might slip into your comfort zone by noting the relative abundance of Dead Sea water during the most recent period of glacial abundance for the northern hemisphere including the years 26,000, 18,000 and continuing even to 10,000 years ago. That imagery holds hope and promise until the vegetation is stripped off and the sand becomes too hot to receive and hold rainfall during the prime growing season.

Relevance To Gulf Of Mexico Dead Zone

You need to realize that the material above contradicts one of the prime tenets of the Ency. Britannica, especially regarding Trade Winds; volume 22 page 372. "The trades, broadly, extend from latitude 30 degrees North to 30 degrees South and thus cover about half of the earth's surface. The great variety of climates included in the trade wind belt ranges from extreme aridity with fog at the eastern margins of the oceans to copious rainfall with tropical storms along the western edges." But then the whole Congo River basin is the visible evidence for the contrarian point of view-that includes the headwaters of the Nile River thousands of miles east of the "aridity and fog at the eastern margin of the ocean . . ." Let's pin at least some our hopes on the cold blast of air from the north as well as the south, contributing at least some-what to the removal of moist air from the Atlantic. I've been on the west end of the Gulf of Mexico. In Arizona the summer monsoon is described as originating in the Gulf of Mexico. And that is in a portion of the state where the Baja California is only 300+ miles away. The next question about the Eastern Mediterranean has a much more complex answer to the question of the source of the cold mass of air that provides the propulsive force which might be induced to drive moist air shoreward in a more seasonally helpful manner. For the answer I'm going to remind you that we need not assert human authority to excess-don't set the norm too narrow-from "Frozen Earth-THE ONCE AND FUTURE STORY OF ICE AGES." p. 149. "But there is one striking feature of the geologic record in these regions that has been linked, indirectly to glaciations; an abundance of coal deposits. The Carboniferous period derives its name from the widespread carbon-rich deposits that occur in this interval of geologic time. They are mostly made up of the remains of spore-bearing plants similar to ferns, plants that lived in low-lying, moist environments often referred to as coal swamps. The organic debris that accumulated in these swamps formed peat deposits; these in turn were eventually transformed into coal. A peculiar feature of the Carboniferous coal deposits in North America and Western Europe is that they are cyclical; beds of coal alternate with marine sedimentary rocks such as limestone or shale in a pattern that is repeated many times over. In places as many as a hundred cycles occur; although it is difficult to determine the amount of time represented by each cycle, in aggregate it is estimated that they span ten million years of deposition, or even more. The plants that were the precursors of the coal grew in **fresh or slightly brackish water, and it is believed that many of the coal deposits formed in low-lying coastal swamps that were periodically inundated with seawater. With each flooding the accumulated peat was buried beneath a layer of ocean sediment; between floodings the fresh water swamps reestablished themselves and new layers of peat accumulated." [Transfer this scenario to the 10,000 square mile dead zone in the gulf of Mexico & you have one natural answer to the increase of dead zones world wide. Other answers will show up as we move along, Author's interjection.] " The pressure and heat of burial eventually transformed the multiple peat layers into the cyclical coal deposits that characterize the Carboniferous." There Sir Richard you have your ice-cold prehistoric carbon sequestration certainty! Is the answer too extreme for you?** It is almost a cinch that the afore-described 10 million year norm is a bit too broad for most of our imaginations. First of all, the bleakness of the Ice Age panorama is engraved a bit too deeply in our minds. We have not reckoned adequately with the change of sea level accompanying the advent of each new ice age. Each studied zone on the continent could have a noticeable drop in temperature caused in part by its sudden **relative increase in elevation courtesy of the sudden drop in sea level. With that in mind it becomes entirely thinkable to have an Ice Age** in the middle of extra warmth.

The First hint of a declining ocean level.

The four or five major physical depressions on the globe, when systematically fed water, could almost harmlessly defeat the expensive hysteria about 'rising oceans. When I say hysteria I'm thinking of the minimum of $20 billion proposed as a partial fix for New Orleans but some of the cost estimates rise frighteningly above the $100 billion mark. It seems almost picayune to mention the $5 billion for upgrading the dikes in the Netherlands and another $5 billion for the water plight of Venice. But who can estimate the cost for the Maldives to purchase a new home on the continent? If the cost for the 360,000 residents of the Maldives to be moved approaches the per person cost of the Bikini residents, (between $1 million and $2 million per person), then the figures rise frighteningly into the mega billions. But establishing wetlands on the lowered oceans like the coal swamps of old would have an incalculable value. I'm betting $25 million that the five major depressed areas of the world can harvest the top 8 inches of the ocean and between storage, oil amelioration, increased evaporation both from the defined zones and from the 70% of the world's now cleaner oceans can produce such a dramatic change that the 'hysteria of rising oceans' will be relegated to the back history portion of the newspapers. The governing 'hysteria' might slip directly to a fever to "protect our newest National Parks—The Oceanic Wetlands." This May only work with the benefit of my father's dust bowl wisdom—"prepare the soil to store and hold the moisture, and pray for RERAINING. But the Mediterranean question which triggered this foregoing material can't be solved by an endless discussion around the question of 'WHAT CAME FIRST? The lack of rain triggered the heat or the heat triggered the lack of rain? We need to exercise Alexander's sword and cut the Gordian knot by providing the additional water to the Qattara and let the rainmaker's magic go to work on this new wealth of evaporated material.

22 degree rise in prehistory contrasted with 2 degrees rise between 1980 and 1993

And now it is time to reintroduce Robert W. Felix, **"Not by Fire But By Ice."** P. 119ff "Changes in sea level coincide with warping of the continents and orogeny (mountain creation) with 'remarkable accuracy,' said J.G. Johnson of Oregon State University. 'The correspondence is so consistent in general, and even in detail, that it must reflect a fundamental relation.' The Arbuckle and Wichita orogenies, the Ancestral Rocky Mt building period and the Palisades, said Johnson, all correlate with sea level fluctuations. (Oros is the Greek word for mountain; geny is from genesis, or creation). Sea level changes do occur in phase with rising and falling land, said LK LK Sloss and Robert Speed of Northwestern University, 'Abrupt localized vertical movements of several kilometers have occurred simultaneously in the past all over the world.' . . . 'I see compelling evidence that orogenesis is cyclic and pulsed,' said Professor S. Warren Carey. 'Expansion waxes to a crescendo, then wanes perhaps to zero before the next wave of expansion. On average, major orogenic activity, including volcanism occurs about every 30 million years. It's linked to our galactic orbit-and to magnetic reversals . . . . ice is not always the culprit. 'The movements are tectonic,' said Sloss and Speed, 'not responses to the weight of ice or mountains? *(why then are most of the world's active volcanoes concentrated within a few miles of the mighty tides of the Pacific Ocean)*? "Same with changing sea levels. The dominant factor, said Anthony Hallam, was more likely to have been changes in the capacity of the oceans due to uplift and subsidence of oceanic ridges and subduction zones in conjunction with continental thickening. Others agree. Sea level changes may result from strong underwater igneous activity (underwater volcanism), said Alexander R. McBirney of the University of Oregon. Sea level changes may be caused by increases and declines in underwater igneous activity, said James P. Kennett of the University of Rhode Island. Aha! Now we're getting somewhere! **Underwater volcanism affects sea levels. And it happened fast! Some sea level rises occurred so fast** that 'great freshwater swamps were flooded within a few hours or days without a gradual transition in salinity,' said Norman D. Newell, in Megacycles; Long-Term Episodicity in Earth and Planetary History." P. 137 more evidence that our present 'global Warming' is a tempest in a teapot, "800 degree plumes of water, from generally small lava flows, are gushing into our seas right now, from the coast of Oregon to the South Pacific. And they're heating our seas! A recent study at the Monterey Bay Aquarium, Research Institute, showed that seawater temperatures are rising. Water temperatures in Monterey Bay have increased almost one degree Centigrade during the last 60 years alone . . . It's not global warming, it's ocean warming, caused by underwater volcanoes. Imagine how it must have been above the massive underwater lava flows of the end-Cretaceous! (Remember, according to Maurice Ewing, all 46,600 miles of our present-day underwater ridge system may have been initiated at the time. 80% of all volcanic eruptions occur underwater, said Steve Hammond, manager of NOAA's Vents Program at the Hatfield Marine Science Center in Newport Oregon. Using NOAA's ratio, and knowing that two-and-a-half million cubic Km of basalt spewed out of the Deccan Traps and Brito-Arctic Flows alone, up to 10 million cubic km.s of basalt could have sizzled into end-Cretaceous seas almost overnight. And every one of those 10 million cubic kilometers would have been unbelievably hot. Up to 2,150 degrees Fahrenheit hot. A planetary-sized hot water heater. That's why ocean temperatures soared. The seas must have boiled, literally boiled, above the underwater ridges. So, we know that thousands, perhaps millions, of cubic miles of basalt spewed into end-Cretaceous seas, initiating our present-day underwater ridge system. We also know the seas had spread far inland, creating thousands of additional square miles of ocean surface, and that ocean temperatures had risen by as much as 22 degrees F. Combine the two, and evaporation had to have increased dramatically." *And did that 22 degree increase provoke the air to accept more than the arbitrary 4% limit for water vapor measured during our short span of scientific interest?*

(Author's interjection) In the oily dry period between 1980 and 1993—the newspaper reported that undersea ocean temps increased by two degrees, threatening and destroying coral reefs-culminating in two feet of hail in Miami and 23 inches of rain in 24 hours in Houston in the early 90's. During that period I was too new to ocean studies to have a standard for undersea ocean temperatures: But two feet of rainfall on land was and still is impressive. Oh! I forgot to mention the billions of dollars of damages done by the '93 floods of Mississippi-Missouri . . . but back to the tale of Robert W. Felix. "Unbelievable amounts of moisture had to have risen into skies already clogged with volcanic debris; skies therefore dark and frigid. Warmer seas and colder skies. a deadly combination. What happens to steam when it cools? It condenses. That moisture had yet to have condensed and fallen to the ground. Rains of Biblical proportions had to have pounded on the newly risen mountains.

Fearfully Sudden Climate Changes In Prehistory! What Now?

Now for the biggest question of all. What would have happened to all of that precipitation if it had been cold at the time? **Snow. Unimaginable amounts of snow."** I may be about to commit the ultimate kind of scholarly abuse in trying to combine the work of Mr. Felix with the fears of James Hansen-but here goes. **Not global warming but ocean warming brings us to the dreaded tipping point. To his credit Mr. Felix points out several** common sense recommendations of things to do and he himself continues to study parts of the earth which may be a bit neglected in the scholarly search for new and relevant data., I heartily recommend reading his book, "Not By Fire But By Ice."—not just once but twice—and study the parts that left you initially stunned.

I have introduced you to Mr. Felix not to emulate the drastic nature of his cheerfully described climatic tipping points in prehistory but to compare and contrast the minimalist nature of the remedies that I may appear to be prescribing-more on this point from Mr. Felix himself.

From chapter 9 p. 160; 'A hail of a mess in Florida,' blared the headline. Two feet of hail fell to the ground in less than half an hour yesterday, the story said, when a freak thunderstorm struck the town of Longwood, Florida. Hail came down so thick and so fast that 'it damaged buildings and snarled traffic.' (Seattle Times, 7 March 1992). Snarled traffic? At two feet deep, that hail stood deeper than the bumpers on their cars. Longwood's traffic wasn't snarled; it was stopped, dead in its tracks.

Figure it out. Longwood's two feet of hail fell to the ground in less than 30 minutes. That's more than 4 feet an hour. Let it hail like that for 24 hours, and you'd have 96 feet of hail.

Or look at Houston, Texas. A relentless rainstorm in October 1994 dumped more than 30 inches of rain on the unprepared city in four days. Add a zero, and you'd have 300 inches-25 feet-of snow. Few roofs in the world could handle that kind of weight."

This factual account of recent history could have been accompanied by the acknowledgement of the afore-mentioned two degree undersea temperature rise between 1980 and 1993-the one that threatened the extinction of so many coral reefs. But when you set the two degree rise beside his projected 22 degree rise in dinosaur extinction times or the equally dramatic extinction of the mammoths 11,500 years ago you may be prepared to accept the pettiness of our present predicament as simply a vaccination meant to prepare against the real thing. (Meaning magnetic reversal leading to more undersea volcanism as well as surface volcanism leading to a glut of airborne vapor as well as particulate-nuclei forming-matter-presto-instant ice age) (return).

We might profitably turn our attention to the 'Presto' part. Again, Robert W. Felix. [His chapters 12 and 13 are more colorful than 'Waterworld,' and more compelling than 'Friday the 13th' or 'Chainsaw Massacre', but his topic is Mammoths—1846 and 1860-the first standing mammoths discovered.] "The first edible mammoth (again, that we know of) was found in the side of a cliff near the Beresovka River in northeast Siberia in August, 1900.

Jutting head-first out of a cliff, with the remainder of its body sealed in a mixture of soil, rock, and ice, the beast was mostly intact, not thawed or rotted away. 'The ice surrounding the carcass was not that of a lake or river,' said a writer identified only by the initials A.S.W., 'but evidently formed from snow.' (NATURE, 30 JULY, 1903)

Once . . . . dissected . . . there were seeds in the giant tummy. The seeds were an important discovery, the Russian scientists said, because they showed that the beast had died in the autumn. (Stomach contents of almost all mammoths ever discovered show that they died in the autumn. <Lister and Bahn,> in Mammoths). Now, if you've ever frozen anything, meat, fish, vegetables-whatever-you know that the quicker you freeze it the better. If you freeze meat slowly, it forms large crystals that burst the cells, thus dehydrating the meat and destroying its flavor. The flesh of many frozen mammoths contain cells that are not burst, indicating that freezing occurred rapidly."

"How . . . . quickfreeze? With nine stories of snow."

Soot Bombs Recommended As Protection For Today!

My purpose in diverting you with a Jason-like chill is not exclusively for entertainment purposes, it is to remind you that any successful carbon sequestration event must have at least a Billion tons of carbon sequestrated as soot bombs to serve as an insurance policy against the catastrophically sudden return of Snowball Earth. On the positive side of the ledger we would have access to adequate amounts of genuinely clean water-possibly with no sex-changing hormones incorporated in it-only having to filter the soot and enhance the delivery system.

Partial documentation of the 60's and 70's blizzard of books and papers announcing the end of temperate climate.

But as I enunciated at the outset of the book-My goal is to achieve balance and so I have shared with you from two books whose major concern is the advance (inevitable)? Of the Next Ice Age. "From Not By Fire But by Ice," heading Ch. 16 page 166. "Previous inter-glacial mild intervals comparable in warmth to the present one lasted only about 10,000 years. The present interglacial has lasted 10,000 years. If it is truly similar to earlier ones-and if man's activities do not alter natural trends-it should be nearly over. James D. Hays-citations include: "1971, "Faunal Extinctions and Reversals of the Earth's Magnetic Field," **Geol. Soc. Am. Bull., Vol 82, p. 2433-2447, Sept. 1971. 1973, "The Ice Age Cometh," Saturday Review of the Sciences, vol. 1, p. 29-32, April 1973. 1976, "Variations in the Earth's Orbit; Pacemaker of the Ice Ages," Science, V. 194, p. 1121-1132, 10 Dec. 1976.0"**

This is just a small portion of the pre-1980 blizzard that I spoke of . . . If I may lapse for a moment into teamster speak-Somebody should stand up and take a bow for averting the inevitable onslaught of ice! By the way, where is he now and did he sign onto the IPCC report?

Has the frozen mammoth with the full belly convinced you that the Ice Age can return suddenly? My gut was convinced by 1979 but my reason has its own pace. I have strong doubts that there are many persons whose head and heart, (gut) are in sync: But I am going to pretend that some of you are now brave enough to walk up to the abyss of (the new ice age) and walk reasonably close to the edge.

Here in skeleton form for comparison and contrast is my recommendation for treading carefully closer to the ice age (reversing global warming)

A. Let Mother Nature have it her way (she will anyway-remember the fly on the wall or the butterfly flapping its wings in China—but when carried to its logical extreme we return to the hunter-gatherer society. XXX-NO.

B. Watch from a safe distance (the moon?) While Mother Nature gushes up gobs of **volcanic goo which initially floods the low lying coastline with rising tides, swelling storms** surges and swelling clouds and endless precipitation—(remember Noah?) XXX NO.

Finding The Middle Way

C. The Middle Way-cooperate with Mother Nature. Recognizing that now as in the past we will have catastrophic events as Southern California is feeling now and New Orleans is feeling the other extreme end of . . . Droughts, floods and wildfires. Some of that may even be exacerbated by the residual presence of unremediated oil on the world's oceans. But I've already told you that this would not be an attack upon the world's fossil fuels. They get credit for temporarily interrupting the normal evaporation and rainfall cycle. The catalytic effect of sun-baked soil and of manufactured hardness, (streets, highways, parking lots, and roofs) deserve a long hard look by those societies that decide to preserve their viability. **C is the reason for this manuscript!**

Let's start this chapter by promoting clean water to the top of our value system. As I've said before this does not need to demote oil-after all combusted oil is the raw material for plant growth-so let's make them co-rulers of the value system. We need a crown prince for this royal family-so I nominate the nuclear-powered ice-breaker and convert him to fresh water harvesting in the only place in the world where there is not only a guaranteed surplus but according to the Milankovitch cycles should continue to be an increasing source of fresh water. And we need a princess for our Royal Family. My nominee is not the fair Indian Princess but is instead the ambitious tree planter from Kenya.

This simple caricature of a theory of course neglects the other central tenet of this book-which is that the civilization induced absence of re-raining is the twin component of my theory. (ONE VILLAIN AT A TIME, PLEASE).

Budyko and Sprengel are not villains

To breathe a little life into this flimsy argument I would like to refer you to an eminent and respected climatologist. Dr. M.I. Budyko (from pp. 181-182 of THE WEATHER CONSPIRACY,). "is associated with the GLOBAL METEROLOGICAL INSTITUTE in Leningrad (now St. Petersburg—book written in 1978) The Budykoyan school . . . The basis of this approach to the climatological problem is Dr. Budyko's 1955 paper entitled, 'THE HEAT BALANCE OF THE EARTH'S SURFACE.' This paper advances the hypothesis that all atmospheric motions are dependent upon the thermodynamic effect of a nonhomogenous distribution of energy on the earth's surface. Though this work originally met with opposition from the world's meteorologists, it is now accepted as a more reasonable basis for developing a successful climate prediction model . . ."

This year's extraordinary amount of Northern Hemisphere snowfall ought to provide a wonderful and frightening proof of the Budykoyan emphasis. The white of snow and the resultant darkening of melting snow into wet earth should help the world's winds out of their recent unusual anemia—and don't forget the increased Infrared absorption of the resulting increase in water vapor. All of this contributes to Mr. Branson's goal of sequestering additional CO_2.

A careful measurement of the springtime drop in airborne CO_2 when compared to the pattern of past years should provide ample evidence that Sprengel's law is at work. In this case the "scarcest resource" is water or (even water vapor—I'll never forget the year when our corn crop came in abundantly on little more than the morning dew). Again, careful measurements of skin cancer rates might show a pause in the growth of this annoying malignancy.

You dear reader are now fully justified in hurling the multiple failures of climate modeling schemes into my face with a "so there"! But I am inclined to cut them considerable slack because they probably have little real world experience with the solar conversion capability of a rained-on field of Nebraska corn or a newly plowed wheat field. All of the minutia of timing of these albedo changes can scarcely be programmed into the largest computer in time to be useful. But I am interested in the big picture.

Unfinished work for scholars—Ocean Phytoplankton

The Nile was once tested with a radioactive isotope to measure the rate of flow beneath the visible Nile. The astounding figures for the rate of flow for the Crypto-Nile have never been corrected or supplemented by new figures for the Post-Aswan Nile—(to my admittedly limited knowledge).

Who has inserted a radioactive isotope above a selected (newly rained on) patch of Sahara Desert? Or for that matter above a lush cornfield in Iowa—or a newly plowed moist earth section in Nebraska? (By the by: My Nebraska kinfolk assure me that there is no more plowing of wheat stubble! Multiplied by one million or more individual choices by farmers it amounts to a significant influence on albedo—probably not factored into the climate scenarios entered into the computer by the modelers . . . Not only is this significant for the rising columns of air that have to return to earth; but also for the trace nutrient value of that same windblown dust when it arrives onto the ocean). The purpose of the radioactive tracer is to measure for the speed of its reappearance in the North Pole fallout zone. Most especially the concern is to know the North Pole transit time of the varying rising columns of air in order to evaluate the timeliness of their contribution to the global air circulation pattern. The comparative speed of rising air off a dry sand or dun-colored desert floor could underscore indelibly the imperative to not interfere with the global wind circulation pattern—that is if we truly desire to avoid the world-wide catastrophe of rising oceans!

New Polar studies will produce measurements of trace plumes of pollutants (nutrients)? It will soon be available via the most comprehensive study yet initiated. I am assuming that no historical comparison is possible with the fallout influence of the late 30's, 40's, and 50's. But some CO_2 charts show an absolute decline and others show a distinct pause in the rate of growth of CO_2 in the period 1962-1964. But the sinking to the 262 foot depth of that ill-fated flight of 5 WW 2 warplanes on the Greenland ice cap should provide a powerful clue as to the influence of airborne (pollutants—nutrients) as they arrive on the ocean surface to accelerate evaporation.

I've read the initial purpose statement of the study. HTTP:WWW.NASA.Gov/MISSION—PAGES/ARCTAS/Airborne—study. html. If this welcome and overdue study has a weakness it is its verbal dependence on studying 'pollution'. I'll repeat Sprengel's Law "Things grow according to the scarcest nutrient . . ." In this case as in most cases—onshore WATER is the scarcest nutrient. In the Ocean the scarcest nutrient is much more complicated, (iron filings may not be sufficient.

I hope I'm wrong and the study expands to consider (positive)? Ocean effects . . . "A second phase of the Arctas campaign takes place this summer from Cold Lake in Alberta, Canada, where flights will focus on measurements of emissions from forest fires (potentially a positive nutrient as oceanic fallout) (And calming influence on increasing acidity of the oceans). Researchers want to know how the impact of naturally occurring fires in the region compare to the pollution (nutrients) associated with human activity at lower latitudes. UNDERSTANDING THE RELATIVE INFLUENCE OF EACH IS IMPORTANT TO PREDICTIONS OF THE ARCTIC'S FUTURE CLIMATE". *(I'M NOT EVEN GOING TO ASK IF THEY HAVE FACTORED IN THE INFLUENCE OF THAT NEWLY DISCOVERED UNDERSEA Arctic Volcano. From: Volcanic eruptions reshape Arctic ocean floor: study-Yahoo! News. 7/11/08 http://news.yahoo.com/s/ afp/2008/sc-afp/sciencegeologyoceansvolcano* And don't forget to note the timing of that (NUTRIENT/POLLUTANT) fallout on the Arctic Ocean. There is a historical period which might provide balancing data for the ocean surface's influence (Translucence versus vegetative murkiness) on the climate. That is the period just prior & including the advent of whaling up to the point where whale harvesting had to be restricted (save the whales). The whales may have acted as giant harvesters multiple cropping the surface layers of the ocean for the nourishing flora in the top most layers of the ocean. This multiple harvesting of the world's oceans by whales maybe one more piece of the missing pieces in the mystery of the growing "dead zone in the Gulf of Mexico." In other words all that rich vegetation decays because nothing (Where are the Whales?) is there to eat it. One other possibility holds a bit more promise. (Later!)

Chapter7

FEAR VS. HYPE. WHICH IS REAL?
VOLCANOES, ATLANTIS, GLACIERS,
OZONE—A 2ND VISIT.

I open this book with a questionable reference to fear mongering—as distinct from the fear mixed with respect even reverence for the potential of our (USA) government. I have attempted to assess and weight a variety of potential hazards to future climate. For example: volcanic dust, nuclear fallout, ozone holes over the South Pole **[Well covered by Susan Solomon]** (to be more fully explored later) but a broader category of ozone shortage could be established by noting the shortage of ozone-creating thunder showers particularly in the period **1980-1993** & even continuing beyond that span of time in other parts of the world particularly **Western Asia and the Sahel.** Ozone holes created by reduced evaporation, ozone holes created by reduced transpiration & oxygen manufacture of surface plants, (this includes the Amazon, the Sahara as well as Antarctic phytoplankton and krill, plus the future effect of cosmic rays in the 2012 GALACTIC PLANE CROSSING, gravitic changes, air mass, ice mass, and tidal forces and going beyond to Planetary alignment). Even though the gravitic influence of planetary alignment amounts to 2% or less than the new-moon full-moon and dawn and sunset influence of the moon and sun on the infant science of earthquake prediction.

Tidal forces bring us to the electrical question and the fear of the North and South Poles swapping ends. And of course not even Robert W. Felix knows what electrical and magnetic changes await on the other side of the GALACTIC PLANE. CONTINENTAL DRIFT OR SHIFT is probably not triggered by comets or meteorites and that probably runs much of the gamut of academic fears.

Even the makers of those charts showing the stratospheric rise of CO_2 are guilty of a little hype since it is up by only a little over 1/3 in a century.

I am going to step over the line with some genuine hype of my own before (hopefully) returning us to sanity with some teamster talk! This chapter is dedicated (minus the following story to a 35 year veteran of the roads and streets of middle America, now retired).

Michael Crichton's Cool Logic

Before I get into my hype, I'd like to ask a question. Could the talk of melting glaciers and rising oceans have been merely hype? I don't have space or time to list all the dissenting views but a little cool logic from Michael Crichton's STATE OF FEAR (p. 423) may help. "There are one hundred sixty thousand glaciers in the world, Ted. About sixty seven thousand have been inventoried, but only a few have been studied with any care. There is mass balance data extending five years or more for only seventy-nine glaciers in the entire world."

If you accept my theory that the oil spills of the late 70's and early 80's provided a temporary hiatus in rainfall that was extended by the subsequent hardening of some soils, you may even be far enough along to be thankful that the spills halted a cooling trend that might have resulted in a precipitate return to THE ICE AGE(S).

OIL OR WATER . . . ONLY ONE IS GOOD TO THE LAST DROP

Acceptable reality is the partly man-caused healing of the sun-locked Great Plains

It may not be compulsory to make 30 or more depth measurements of one hundred sixty thousand glaciers in order to lay the 'rising oceans questions' to rest. Our focus should be on the techniques to renew and soften the multiple at-risk soils of the earth.

Now for my particular brand of hype!!!

When late night bull sessions at college turn wonderingly to the myth (circa 1958) of Atlantis—and disappearing continents and civilizations, they very rarely turn to factual wonderment about the missing portion of human history locked into the below sea level zone from sea level down to 600 feet below. That portion of pre-history is still best captured by the various Flood (Noah et.al) legends.

Only two inexplicable facts are illuminated by this admittedly wild-hare theory as well as the murky linking of Atlantis-Antarctica. The first mystery is Lake Missoula and the unsubstantiated theory that its sudden emptying was triggered by the breaking up of a gigantic ice dam on the **South** side of the North American Ice Age. Why should Ice on the sunny side of the ice field be the last to break up? Admittedly the recent radar aided discovery of large freshwater lakes underneath Antarctic ice (especially the threatened rapid recession or drainage of same) strengthens that theory. Let me continue with the previously mentioned mystery of the sudden freezing disappearance of the Siberian mammoth. Suppose that the pile of ice for the last ice age centered on the nation of Canada actually represented the North Pole and the subsequent motions of Continents (southward for the Americas and northward for Siberia). I have no notion of the contrariness of this idea for the proponents of gradual continental drift (plate tectonics) but I may expect to hear soon in words of less than four syllables. The point of this improbable scenario should be made obvious. *Nobody alive can comprehend the stresses affecting* *continental drift which might impel a whole continent to motion when the forces of the surrounding seas are lessened by 400 to 600 feet and the weight of the missing 400 feet of water is concentrated on North America.*

The evidence of the many volcanic chains of islands (Hawaii, Aleutians even Yellow Stone Hot Spots) successfully rebut the fears of sudden continental shift implied here but bear with me a bit.

The above is of course a deliberate descent into fear mongering. Only one sentence may be useful in future deliberations-now for visible reality. I posed the question (1st Tuesday of March 2008) to a breakfast of retired teamsters. What major fear arises in your mind as a result of the publicity surrounding the theme of **"Global Warming?"** The first answer came instantly from the biggest guy at the table: "Fear of Drowning"! We were sitting approximately 1,000 feet above sea level at the time. There were a couple of other answers-maybe fear of larger AC bills and of course larger hurricanes but fear of drought (with the Missouri river running close by) was not mentioned.

Colloquial Reality!

But the big answer was given with some emotion and rang emotional chimes around the table. "I'm afraid of our government" were the decisive words. I'm ready to take (some of) the blame for the ideas that follow that lead. The guy who said it admitted that he drove his pickup about 7,000 miles per year and the government regularly hoists gigantic loads into orbit (making temporary holes in the ozone layer) without apology and (here incredulity pervades his voice) "they want me to carry the load of guilt"! It is now the 1st of Febr. 2010 and the same teamster who was critical of our government was caught driving a brand new four-wheel drive pickup and his immortal quote runs like this! "I quit believing in global warming, 'cause this winter made me spend a lot of money to buy a four wheel drive." I should apologize for dawdling on the manuscript this long: But his common sense almost obliterates the reality of the explosive forces of nature that can arise from an ocean that somehow is required to accept and store excess quantities of heat. That means either solar by way of ocean sealing or volcanic, caused in part by increasing gravitic flux, or any of several almost wholly natural means. But I mustn't neglect the third almost wholly man-caused reason. When the land becomes sun-locked and refuses by reason of surface heating to accept precipitation, it then deserves the concerted efforts of the most thoughtful minds and actions much as the Dust Bowl received the attentions of both Government and farmers in the 30's.

A minority view, a 2nd kindly comparative look at the dirty 30's

In a bit more reflective vein the government of the 30's withstood unemployment figures around 30% and they still built highways, planted trees and windbreaks, built dams and levees, counseled enormously productive conservation measures and built the greatest military machine the world has ever seen. Since that time we've had the advent of political correctness, diversion of the Corps of Engineers to building hospitals and roads. The roads are full to exasperation and decay with an onslaught of male and female wage slaves, and the state has moved massively into the broken families that result. The work ethic that once sustained America has shifted to the sport competition world and the broken bodies that result continue to finance the exponential growth of the medico-pharmacy types that prey on the athlete's hyper emotionally driven state. The medical system's claim to fame (increased lifespan) is countered by the world's ranking of U.S. health system (currently 42nd or lower) and I can only wonder aloud if the eugenically aborted 48,000,000 babies were averaged into that claim? Even Doctor's life spans are questionable-certainly they fall short of Dr. Albert Schweitzer's 95 years most of it in a not very welcoming climate.

That's just off the top of my head! Now for the serious stuff!

Most of the foregoing material has attempted to reach far enough into the distant past to give a safe overview of possible changes that may occur in the course of nature. In a bit I'll quote from an acknowledged expert on fear, but for now let's recognize that I'm about to step into the quicksand of present reality. The question is "What can we do to amend the mistakes of the past 85 years recognizing that those 'mistakes' were simply put an inoculation or preparation for the decisions to be made in the present reality. Future rational plans cannot fail to be aided by the following excerpts from Michael Crichton's **"State of Fear"**. Since redwood trees (Sequoias) might become the centerpiece of an extensive carbon sequestration effort and since Yellowstone Park might become the center for insurance plans to avoid a **"Precipitate"** return to an ice age, we should welcome the following.

M. Crichton, Yellowstone, the Sequoias—water and fire in prehistory.

From **Crichton's "State of Fear"** pg. 402 ff, "The threat of abrupt climate change,' he said 'is so devastating for mankind, and for all alike on this planet, that conferences in Los Angeles starting tomorrow, where scientists will discuss what we can do to mitigate this terrible threat. But if we do nothing, catastrophe looms. And these mighty, magnificent trees will be a memory, a postcard from the past, a snapshot of man's inhumanity to the natural world . . . . and only we can stop it.'

'That . . . was all bullshit'. Jennifer said . . . . 'The whole speech . . . And they've managed to survive forest fires? Hardly, they're dependent on fires, because that's how they reproduce. Redwoods have tough seeds that only burst open in the heat of a fire. Fires are essential for the health of the redwood forest' . . . How long do you think your primeval forest has looked the way it does now?

'Obviously, [not] for hundreds of thousands of year.'

'Human beings were here for many thousands of years before these forests ever appeared . . . Let me lay it out for you'.

Twenty thousand years ago, the Ice Age glaciers receded from California, gouging out Yosemite Valley and other beauty spots as they left. As the ice walls withdrew, they left behind a gunky damp plain with lots of lakes fed by the melting glaciers, but no vegetation at all. It was basically wet sand.

After a few thousand years, the land dried as the glaciers continued to move farther north. This region of California became arctic tundra, with tall grasses supporting little animals, like mice and squirrels. Human beings have arrived here by then, hunting the small animals and setting fires. 'Okay so far? Jennifer said, 'No primeval forests yet'.

I'm listening, 'Ted growled . . . .

She continued. 'At first, arctic grasses and shrubs were the only plants that could take hold in the barren glacial soil. But when they died they decomposed, and over thousands of years a layer of topsoil built up. And that initiated a sequence of plant colonization that was basically the same everywhere in post-glacial North America.

First lodgepole pine comes in. That's around fourteen thousand years ago. Later It's joined by spruce, hemlock, and alder—trees that are hardy, but can't be first. These trees constitute the real 'primary forest and they dominated this landscape for the next four thousand years. Then the climate changed. It got much warmer, and all the glaciers in California melted. There were no glaciers at all in California back then. It was warm and dry, there were lots of fires, and the primary forest burned. It was replaced by the plains-type vegetation of oak trees and prairie herbs. And a few Douglas fir trees, but not many, because the climate was too dry for fir trees.

Then, around six thousand years ago, the climate changed again. It became wetter, and the Douglas fir, hemlock and cedar moved in and took over the land, creating the great closed canopy forests that you see now. But someone might refer to these fir trees as a pest plant—an oversized weed—that invaded the landscape, crowding out the native plants that had been there before them. Because these big canopy forests made the ground too dark for other trees to survive. And since there were frequent fires, the closed-canopy forests were able to spread like mad. So they're not timeless, Ted. They're merely the last in line'.

Bradley snorted. 'They're still six thousand years old, for God's sake.'

But Jennifer was relentless. 'Not true,' she said. Scientists have shown that the forests continuously changed their composition. Each thousand year period was different from the one before it. The forests changed constantly, Ted. And then of course, there were the Indians.' . . . . The Indians were expert observers of the natural world, so they realized that **old growth forests sucked.** Those forests may look impressive but they're dead landscapes for game. So the Indians set fires, making sure the forests burned down periodically. They made sure there were only islands of old growth forest in the midst of plains and meadows. The forests that the first Europeans saw were hardly primeval. They were cultivated, Ted. And it's not surprising that one hundred fifty years ago, there was less old-growth forest than there is today. The Indians were realists. Today it's all romantic mythology' . . ."

The projected use for giant Sequoias as vertical carbon sequestration units keeps us on the relatively safe topic of righting the 4,000 to 5,000 year (+) grievance of the desertification of the Sahara. Not only would they contribute to carbon storage but their height would contribute to the necessary turbulence for the rainmaker's skills to come into play. The site is of course the Qattara Depression. The plan (such as it is) is to ring or border the man-made (lake, sea, reservoir) at the predetermined height which maximizes both power production and evapo-transpiration. My advisor counsels **Tulip Poplars** as the necessary precursor to the establishment of a successful stand of Sequoias. The foregoing material prepares us to accept that advice and cautions that Nature-Caused-Change will continue and probably prevail

Rain-making after Charles Mallory Hatfield.

Now it's time to gratefully revisit Charles Mallory Hatfield. His 1916. San Diego flood may continue to provide useful tools for Southern

California and its water problems. For the Qatar depression we need to update ourselves on the modern rainmaking skills. From the April edition 2008 of Popular Mechanics, p. 60. The article is entitled, who'll start the rain? By Elizabeth Svoboda.

From the cockpit of a Cessna 340-A 10,000 feet over North Dakota, I watch a bucolic conveyor belt of farming towns scroll below. Then, the whole world goes grayish white.

'We're just popping through the edge of this cloud bank,' says Hans Ahlness, a pilot and vice president of the Fargo-based company Weather Modification, which specializes in teasing rain from the sky. Ahlness points to a mushroom formation a few hundred feet away. 'See how that cloud's got a cap on top? That means it's growing fast. That's what we're looking for.' As we approach the gray cluster, the plane bounces up and down, like an elevator car on a slack cable. I swallow, fervently willing my breakfast to stay in my stomach. 'We're getting a little bit of an updraft from the cloud', Ahlness says calmly pointing to the altitude gauge on the instrument panel. 'See how were maintaining speed and still climbing?'

One of the cardinal skills taught in flight training is how to steer clear of burgeoning storms, which tossed two seater planes around like balsa wood gliders, and test the mettle of even experienced pilots. But Ahlness and his fellow cloud seeders fly directly into the roiling depths, firing dozens of foot-long flares that disperse a cocktail of salty substances as they burn up. The salt forms millions of ice nuclei that attract droplets of water. Eventually, the drops grow heavy enough to fall out of the sky as rain. 'Our mission is to make inefficient clouds more efficient with aerosols that are lacking in nature,' says Bruce Boe, Weather Modification's director of meteorology

Humans have been trying to change the weather since the earliest rain dances; showers produced by the first modern rainmaking experiments in the 1940s and 50s seemed more symbolic than practical. But the prospect of conjuring precipitation from stingy clouds has lost its mad scientist overtones in recent years. Nearly half of the contiguous United States experienced drought in 2007, and scientists predict that water shortages will worsen with global warming. For many regions, engineering the weather is now viewed as an absolute necessity.

State and county governments have begun to fund weather modification gurus with increasing urgency: The Wyoming water development commission allocated $9 million for a five-year cloud seeding research program. Similar efforts are underway in North Dakota, California, Utah, Colorado, and Texas. Even Congress has gotten in on the act: Last year, Texas Sen. Kay Bailey Hutchison introduced a bill to establish a national weather mitigation policy.

Rainmaking as modified by the concept of RERAINING!

I'm going to insert my own guidance for the 'national weather mitigation policy.' When the concept of 'RERAINING' is fully established in the mind of the legislator, the areas for cloud seeding practically sit-up and beg to be noticed. That should mean that seeding should occur at altitude and at a location where prevailing and seasonal winds might augment further rainfall downwind from the original site. If we really get good at the application of this idea to the dry zones of California and the arid Southwest we might eventually receive an "A" grade in the category of 'Rain Harvesting.'
Back to **Popular Mechanics p. 62.**

'I'm just going to pass through this little cloud right here', Ahlness calls out, grinning. 'Hang on!'The Cessna sweeps and dips like an agitated bird. Beneath the plane's wings are instruments that count precipitation droplets, sample aerosols, sense wind flow and measure the liquid content of clouds. While flares haven't changed much since the 1960s—salt-based silver iodide is still the seeding agent of choice for most firms—the addition of remote-sensing instrumentation is relatively new, and increasingly common. Fed into sophisticated weather simulation models, this data gives scientists a much more detailed portrait of cloud conditions— including factors like the size and temperature of existing droplets, the presence or absence of ice crystals and the concentrations of pollutants—and predicts how successful a particular type of seeding will be Silver iodide, for instance, works best on supercooled clouds made of water droplets colder than 23°F.

'It used to be that pilots would just take off and find the strongest updraft to decide where to seed,' says Darren Langerude, director of North Dakota's atmospheric resource board. 'Now researchers are able to tell you things like," you don't want to seed the main updraft, you want to seed another updraft on the flank of the cloud formation.' *Still, despite high tech refinements, some scientists remain unsure whether the modifiers' strategies are any more effective than burnt offerings to the rain gods. This skepticism dates back decades: Federal funding for cloud seeding programs dwindled to almost 0 in the 1990s, when studies that gauge the effectiveness of different rainmaking strategies yielded mixed findings. [Author's interjection. What is needed is a triple philosophy of seeding both onshore nearshore and on the ocean. In other words in North Dakota there are limits as to how much blood you can squeeze out of a turnip.]*

"Current research indicates that, on average, cloud seeding measures boost precipitation in North Dakota by 4 to 10% annually—and up to 20% across the nation—but because weather conditions vary so much from year to year, seeding programs don't often obtain consistent results. A 2003 National Academy of Sciences report concludes: 'there is ample evidence that seeding a cloud with a chemical agent can modify the club's development and precipitation. However, scientists are still unable to confirm that these induced changes result in verifiable, repeatable changes in rainfall.' When Bill Cotton, an atmospheric scientist at Colorado State University, developed a computer model to evaluate seeding measures in Colorado, he came up empty. 'The model didn't show much of the difference between seeding and no seeding,' he says.

"Most cloud seeders have a healthy understanding of the limitations of their trade. 'A 4 to 10% increase in rainfall isn't a tremendous amount,' Langerud says, 'but it could translate into an inch of additional rain during the growing season.' North Dakota counties that have signed on for seeding enjoy roughly $8 million in increased agricultural production each year, with an annual investment of about $700,000. These returns, according to Langerud, make seeding a more attractive drought busting measure for cash strapped states than traditional fixes such as building new dams and reservoirs. 'It's like,' I'll give you a dollar, you give me 10 back,'" he says. 'That seems like a pretty good deal to me.'"

My gut feeling based on what I know of the Science (Art) of reraining is that the 10% increase in rainfall, is devalued in the statistics because the effects of seeding, properly done, affect a much wider area than the subscribed zone, and consequently the downwind rainfall (which the seeding caused) becomes part of the standard which may suggest that nothing of much statistical import took place with the initial seeding, Additionally the welcome presence of the clouds, even without the rain, modifies in a welcome way the merciless beating of day after day of endless sunshine. We'll have a recommended time and place for seeding when we deal with the arid Southwest!

Nevertheless, 10 to 1 payback is still pretty good.—Back to the article!

"On this particular August day, our mission into the upper troposphere seemed doomed. In the old days, pilots like Ahlness would have gone with their gut instincts. But modeling data from the ground crew indicates these clouds aren't big or droplet packed enough to make firing flares worthwhile. 'Every system is different,' Ahlness says as we head back toward the ground. Despite the false starts and the ongoing debate over clouds seeding's merits, Ahlness and his fellow pilots remain true believers, willing to devote their careers—and even their free time—to the cause . . .".

"Like emergency room doctors, Weather Modification's wingmen must keep their pagers on 24/7 in case they are called on for middle-of-the-night missions. They love to grouse about the inconvenience, but it"s obvious they live for the rush. 'Look—THERE'S A CELL OVER THERE THAT'S JUST GETTING GOING,' Ahlness says with a kid's unfettered excitement. As soon as the burgeoning fronts around us converge into another perfect storm, he'll take to the skies again."

NOW AFTER NECESSARY INTERRUPTIONS—THE SAHARA PROJECT QATTARA ENLARGED.

'The Qattara/Sahara project in its bare bones steps outside of the time constraints imposed by Mr. Branson's contest but one hundred years to rectify a five thousand year old crime makes it a worthwhile investment. In its minimal realization one could hope for a narrowing of the **Dead zone of the expanding Sahara**— most notably covering the 60's, 70's and 80's. More to come in my revision of the 100 year old **Qattara Depression Project.**

Jennifer continues . . . ."The threat of global warming,' she said, 'is essentially nonexistent. Even if it were a real phenomenon, it would probably result in a net benefit to most of the world."

Above and below: praise and a limited rebuke to the stand-pat position

There in a nutshell is the position of the non-greens and there is enough truth in it to allow the stand pat position to withstand all but the most carefully formed counter-arguments. Other than the progressive decimation of the Sahel, the Trillion dollar disturbance in Western Asia, the twelve-to-twenty million surge of brown's across USA's southern border, (who are in most cases merely searching for some healthy green.) Then, there is the trifling matter of unprecedented storm surge in New Orleans, Bangladesh, and of course we might include a small portion of Tsunami damages that may have been aggravated by the revision of global atmospheric water content downward from 4%, 2%, or 3% to 1% or 2% following the major oil releases of the late 70's and early 80's. The only goal that can withstand the debilitating glare of the stand-patters is this; To bring the Life-giving, rain-enhancing descent of the Jet Stream far enough south in the Northern Hemisphere to bring its moist blessings to the equatorial regions that have been given short shrift. And of course that cannot succeed in a permanent way unless societies value that rain enough to trap it or delay it & return it by normal land based transpiration to the atmosphere lest it return quickly to the relative anonymity of the ocean. On the land this moisture produces reduced albedo (increased solar conversion by way of increased infrared absorption at the top of the atmosphere.

Another factual side to the rehab job for the Sahara—Sahel

How much extra ozone is created on the rising columns of Amazon rain forest oxygenated air? The answer that I didn't have courage enough to drum into your head on page five is that you don't need any ozone on a cloudy column of rising water vapor because the ultraviolet penetration is stymied by the clouds. But the question is still valid in a deeper sense because lightning strikes above the clouds do create ozone even if its destination is unknown. ((My Interjection

The question belongs everywhere and nowhere) The moisture produces increased surface conversion via color changes both in the water and by the darkening effect of surface organics.) And let's not neglect the global effect of the adiabatic rising air currents that can add a plus to wind currents in a timely welcome global fashion. *   *   * It is said in the Reader's Digest book "STRANGE STORIES AMAZING FACTS" published in 1976 "that the world's lakes & rivers contain approximately 55,000 cubic miles of water. The same article claims the Sahara Desert has 150,000 cubic miles of water beneath it in underground reservoirs. When this factoid is set into print it is obvious to me that we are still in the hysterical jargon and need to move onto firmer ground in the debate of rising Oceans. From the same book RIVERS and Streams account for only 0.001% with 300 cubic miles in the rivers. Other lake water brings the total of the world's lakes & rivers to approximately 55,000 cubic miles of water. A percentage of that retained on land would make an appreciable dent into the hysteria of rising oceans. Under specific plans I will **remind you of the altered job descriptions for the** Corps of Engineers as well as civilization's need to retain water in specific areas of the planet.

I should mention the dramatic albedo effects of this winter's inundation of the Northern Hemisphere with bright snow. 50 years ago this would have been regarded as a possibly normal winter. But our expectations have been reconditioned by a succession of dryer than normal winters. I mention this in order to prepare the climate change scholars among you to note the continuation effect of abundant springtime moisture and its consequent spring & summer effects upon global air circulation patterns. Some of those effects should show up in an increasing profit margin for the giant windmills now proliferating across the world. **This material was written before the winter of '09**

In the continuing debate of natural vs. man-caused: I'm inclined to cut both sides some slack in favor of the concept of adequate on shore precipitation which in its northern reaches may mean slowed runoff and increased absorption.

Chapter 8

YELLOWSTONE, A HISTORICAL REVISIT, CRICHTON AGAIN!

But I'm still concerned about criticizing the government that I love and admire. Michael Crichton does it best—YELLOWSTONE_ from STATE OF FEAR, p. 484 ff. "Yellowstone Park, ... was the first wilderness to be set aside as a natural preserve anywhere in the world. The region around the Yellowstone River in Wyoming had long been recognized for its wondrous scenic beauty. Lewis and Clark sang its praises. Artists like Bierstadt and Moran painted it. And the new Northern Pacific Railroad wanted a scenic attraction to draw tourists west. So in 1872, in part because of railroad pressure, President Ulysses Grant set aside two million acres and created Yellowstone National Park.

There was only one problem, unacknowledged then and later. No one had any experience trying to preserve wilderness. (What about Sherwood Forest?) There had never been any need to do it before. And it was assumed to be much easier than it proved to be.

When Theodore Roosevelt visited the park in 1903, he saw a landscape teeming with game. There were thousands of elk, buffalo, black bear, deer, mountain lions, grizzlies, coyotes, wolves, and bighorn sheep. By that time there were rules in place to keep things as they were. Soon after that, the Park Service was formed, a new bureaucracy whose sole job was to maintain the park in its original condition.

Yet within 10 years, '[03-1913?]' the teeming landscape that Roosevelt saw was gone forever. And the reason for this was the park managers—charged with keeping the park in pristine condition—had taken a series of steps that they thought were in the best interest of preserving the park and its animals. But they were wrong.

'Well,' Bradley said. 'our knowledge has increased with time.' . . .

'No, it hasn't' Kenner said. 'That's my point. It's a perpetual claim that we know more today, and it's not borne out by what actually happened.'

Which was this: the early park managers mistakenly believed that elk were about to become extinct. So they tried to increase the elk herds within the park by eliminating predators. To that end they shot and poisoned all the wolves in the park. And they prohibited Indians from hunting in the park, though Yellowstone was a traditional hunting ground.

Protected, the elk herds exploded, and ate so much of certain trees and grasses that the ecology of the area began to change. The elk ate the trees that the beavers used to make dams so the beavers vanished. That was when the managers discovered beavers were vital to the overall water management of the region.

When the beavers disappeared, the meadows dried up; the trout and otter vanished; soil erosion increased; and the park ecology changed even further.

By the 1920's it had become abundantly clear there were too many elk, so the rangers began to shoot them by the thousands. But the change in plant ecology seemed to be permanent; the old mix of trees and grasses did not return.

It also became increasingly clear that the Indian hunters of old had exerted a valuable ecological influence on the park lands by keeping down the numbers of elk, moose, and bison. This belated recognition came as part of a more general understanding that native Americans had strongly shaped the 'untouched wilderness' that the first white men saw—or thought they were seeing—when they first arrived in the New World. The 'untouched wilderness' was nothing of the sort. Human beings on the North American continent had exerted a huge influence on the environment for thousands of years—burning plains grasses, modifying forests, thinning specific animal populations, and hunting others to extinction." (And don't forget to add the world-wide influence of animal migrations: The Bison in America or Zebra, wildebeest, or elephant in Africa. Their annual dry weather migrations with the colossal clouds of dust they raise have its own timely micro nutrient addition to the ocean that ultimately results in additional on shore precipitation and the cooling effect of more abundant clouds).

"In retrospect, the rule forbidding Indians from hunting was seen as a mistake. But it was just one of many mistakes that continued to be made in an unbroken stream by park managers. Grizzlies were protected, and then killed off. Wolves were killed off, and then brought back. Animal research involving field study and radio collars was halted, and then resumed after certain species were declared endangered. A policy of fire prevention was instituted, with no understanding of the regenerative effects of fire. When the policy was finally reversed, thousands of acres burned so hotly that the ground was sterilized, and the forests did not grow back without reseeding. Rainbow trout were introduced in the 1970's, soon killing off the native cutthroat species.

And on and on and on.

'So what you have,' Kenner said, 'is a history of ignorant, incompetent, and disastrously intrusive intervention, followed by attempts to repair the intervention, followed by attempts to repair the damage caused by the repairs, as dramatic as any oil spill or toxic dumps. Except in this case there is no evil corporation or fossil fuel economy to blame. This disaster was caused by environmentalists charged with protecting the wilderness, who made one dreadful mistake after another—and along the way, proved how little they understood the environment they intended to protect.'

'This is absurd.' Bradley said. 'To preserve a wilderness, you just preserve it. You leave it alone and let the balance of nature take over. That's all that is required.'

'Absolutely wrong.' Kenner said. 'Passive protection—leaving things alone—doesn't preserve the status in a wilderness any more than it does in your backyard. The world is alive Ted. Things are constantly in flux. Species are winning, losing, rising, falling, taking over, being pushed back. Merely setting aside wilderness doesn't freeze it in its present state; any more than locking your children in a room will prevent them from growing up. Ours is a changing world, and if you want to preserve a piece of land in a particular state, you have to decide what that state is, and then actively, even aggressively, manage it.'"

Manage it! Yellowstone as accidental victim of teleconnections from the South and West toward L.A. even the Eastern Pacific.

ENOUGH (for now) of that reproof of environmental greens: Let's lighten the criticism a bit. Remember the period 1903-1913 the downturn time for Yellowstone. From ENCYCLOPEDIA BRITANNICA circa 1965 vol. 14 pg. 320 NATURAL RESOURCES . . ."Associated with Los Angeles as much as oranges or motion pictures, however, has been large-scale oil production. A natural bituminous pitch, oozing out of the earth and called BREA, was used by the Indians and early Spaniards to coat the roofs of their dwellings. After crude experiments to distill this product, in the 1850's and 1860's, there followed the colorful era of Edward L. Doheny who began the first widespread development of the gas and oil fields surrounding the city in the 1890-'s SEVERAL OIL COMPANIES WERE FOUNDED WITHIN THE FIRST DECADES OF THE 20TH CENTURY AND VAST QUANTITIES OF OIL WERE TAKEN FROM NEWLY DISCOVERED SIGNAL HILL, SANTA FE SPRINGS, AND HUNTINGTON BEACH. THE PRODUCTION OF THESE THREE FIELDS WAS SO GREAT THAT THEY FREQUENTLY UPSET THE NATIONAL OIL PRICE, STORAGE,

AND DISTRIBUTION STRUCTURE THROUGHOUT THE 1920'S AND 1930'S. HUNDREDS OF MILLIONS OF BARRELS WERE PRODUCED BY THESE AND OTHER LOS ANGELES WELLS . . ."

The dates do not jibe with courtroom precision but on a **1958** visit to five northern California cities, (a week at each, including the fruited central valley) I found widespread resentment directed at Los Angeles—whom they held responsible for their ongoing water availability plight. Need I add that Yellowstone Park lies in the northeasterly position from Los Angeles and of course considerably closer was Northern California who felt their water had been robbed from them; hence less of or no re-raining in a **North Easterly** direction. A great circle route roughly identical in shape to the fallout charts from the nuclear testing which would begin **40** years later—a few hundred miles to the east. IF ENOUGH OIL WAS SPILLED IN LOS ANGELES BASIN AS WAS SPILLED EVERYWHERE ELSE IN THIS ERA—*IT COULD HAVE AFFECTED RAINFALL PATTERNS NOT ONLY FOR ANGRY NORTHERN CALIFORNIANS, BUT EVEN AS FAR AS YELLOWSTONE PARK. By the way there are new figures out about natural oil seepage in the Los Angeles bay area. The claim was made that the seepage equaled the total of the Exxon Valdez spill in any four year period. I have no notion of the claim's validity, but a friend of mine who grew up there recently returned for a visit and came back claiming an unpleasant oily stench, rolling in off the ocean.*

Schweitzer and the Gabon another look for balancing climate norm!

Lest I be accused of joining the BOF crowd (BLAME OIL FIRST) I have selected some material from Albert Schweitzer's biography, GENIUS IN THE JUNGLE, Joseph Gollomb, Vanguard Press 1949, 16th printing. The speech is the preparation that Schweitzer gives to his intended prior to their June 18, 1912 wedding. Safely before any major equatorial ocean corrupting influences could be attributed to the oil companies.

"Helene Breslau was the only one to whom he confided that he decided to spend the next eight years in the study of medicine, and that then he would go to the Gabon to cure the sick there.

She said she would wait for him to complete his course, and then she would go with him . . . .

And he pointed out the hardships that would be hers in the Gabon. It was known as 'the most unhealthful spot on earth'. Only forty miles from the equator, it had a killing climate; a deadly sun and only two seasons, TORRENTIAL RAIN AND A DRY SEASON with the heat of a steam room. Everything disintegrated in the dank heat there, clothing, furniture, even buildings. What the climate did not destroy, termites and a host of voracious insects did. It was a country where cattle could not be raised because of the tsetse fly, ... carrier of sleeping sickness to beast and man. It was inhabited by the most primitive tribes on earth, most of them cannibals. They were afflicted with some of the worst diseases the whites could contribute, and they had a whole host of their own; raging, highly contagious fevers, virulent dysentery that laid low whole communities, heart ailments, skin diseases that ate deep into the tissues, limbs swollen to elephantine size, leprosy, insanity. And famine swept often over the land.

Did he have the right, he demanded, to expose her to such a malignant world? She thought it all over seriously, and then she smiled.I will take a training course in nursing,' she said.'Then you won't be able to get along without me' . . .

'Paradox though it may seem,' Schweitzer wrote, 'nowhere is it easier to starve then amidst the luxurious vegetation of the game-haunted forest of Equatorial Africa.'

For, lavish though the region was with jungle growth, it was not generous with food for human beings. Practically all the citrus fruits, bananas, yams, potatoes, and other fruits and vegetables had been introduced along the Ogowe' by whites. The natives had learned to grow bananas, plantains, and other foods, but cultivation has been crude and beset with difficulties. The banana rapidly exhausts the soil, and the natives have to clear the jungle for new plantations every few years. Elephant herds often raid them, and whatever they don't eat they destroy.

Rains often fail the natives, and drought brings famine . . ."

This data which is not quite so ancient is included for more than the demonstration of a non-oil caused drought and flood cycle. The chart of the following page conveys a precise picture of the temperature and rainfall patterns of the equatorial country with close proximity to the Congo River. I include it because Schweiter's times there mostly predates the influence of equatorial oil spillage of the abundance of fossil fuels. The only other significant weather reference is on p. 243 where rainfall in 1945 interrupts the brief growing season to bring famine.

"The food situation had been bad in the Gabon since 1945. Rain had fallen in the dry season of that year, preventing the natives from growing a crop of the two staples, bananas and tapioca." I needed that to reprove my latent bias in favor of almost any kind of onshore precipitation . . . . a logical bias triggered by my youthful exposure to semiarid Nebraska soil.

The fact of Gabon's midsummer rains and subsequent famine, coupled with my extensive collection of dust stories—which may have reached their volumetric peak in 1945—should help to convince some lingering doubters that man can indeed influence weather, certainly, and (climate?—Maybe?)

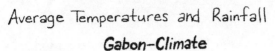

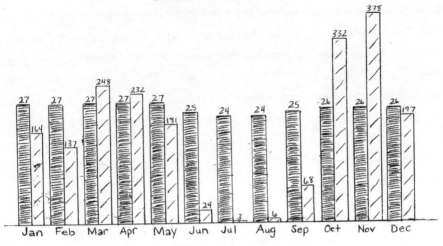

☐ Temperatures °C in Libreville

▨ Rainfall (mm) in Libreville

(Average values for each month)

drawn from Genius in the Jungle, by Joseph Gollomb

The above chart is fairly complete except for any indication of the length of the charted period.

Partly answered questions about the huge monthly variability of rainfall in equatorial Gabon.

Average temperature varies within a 3 degree Celsius range. The rainfall numbers vary spectacularly from 3mm in July to 378mm in November. I cannot explain these numbers with dogmatic certainty because I don't have at my fingertips right now the timing of the annual

149

springtime flood of the Amazon; or Congo, or Nile or Niger. But, I'll try anyway. The June, July, August and even September's moderate deficit is explained by the (approx) May incursion of (cooler?) Amazon floodwaters into the Atlantic. *Or perhaps better explained by the powerful sucking updraft of the spring rains on the West side of the Atlantic—Brazil.* The rainfall upturn in October and November occurs because of the Vegetative murkiness spreading from the Amazon—200 miles or more out into the Atlantic. That leaves only the March and April mini-peaks to be traced to the six month heating of South Pole waters, and their subsequent (miraculous?) transfer Northward to their needy equatorial destinations.

Chapter 9

INTRODUCTION OF MICHAEL COLLIER'S FLOODS, DROUGHTS AND CLIMATE CHANGE.

To my best knowledge the ocean source of onshore precipitation has never been accurately pinpointed in a systematic way except for the El Nino influence & even this data incorporates huge mysteries. We need to know that in order to economically revegetate the Sahara. But theories of this sort simply confuse me. It's a whole lot less confusing to hold the mental picture of massive lake effect snowfall on the East end of Lake Erie where the cold dry northwest wind picks up the warmer water and dumps it on Buffalo at the far end of lake Erie. Or perhaps we can be simply illuminated by the rainiest area in the world. From p. 91 floods 2nd par "Floods Drought and Climate Change" Michael Collier pub. 2002 Univ. of Ariz. Press, Tucson. "As the air over India and the Himalayas heats during spring, it rises, drawing moisture from the surrounding oceans. Moist air rising over hot rock: This is the perfect setup for convective thunderstorms of epic proportion. This rising air is the source of the Asian monsoon. No other place on the earth boasts a greater rainy season. Cherrapunji, in the Assam Hills of India has received 25m of rain (yes, 82 feet) in a single year, arriving mostly during the four-month monsoon. The monsoon occurs seasonally rather than year-round because it is tied to the northward summer migration of the ITCZ." Reality for me is that these rains don't seem to rely on the confusing presence of nucleating particulate matter or selected hot spots of the ocean as exampled by the all time record rainfall (1861 East India culminating in 36 feet of rainfall in the month of July . . . . ITCZ.

We are almost ready to consider the Qattara Depression Project. But before I begin I must confess my own inadequacy for the humanitarian side of the project. The multitudinous kinds of timing required for the human success of this project involves more knowledge of the peaks and valleys of the three rivers—Amazon, Congo and Nile—than I have thus far demonstrated. *And more especially for the peaks and valleys of Solar and Cosmic Ray activity, which is why I will gladly turn the problem of rehabbing the rainfall side of the Sahara to Nasa, primarily because they have thought worthwhile thoughts about developing waterless planets, but also because they have the most accurate up-to-date and comprehensive measurements of earth's climate.* But I think I have demonstrated that the interrelationship begins at least as far south as Antarctica.

Precipitation pattern (for some climatologists) is irrevocably altered. For evidence we have only to note the first ever South Atlantic hurricane. Forestalling an epidemic of South Atlantic hurricanes similar to the North and Middle Atlantic hurricanes should be a legitimate goal of those of us who would alleviate man-caused climate changes.

This homespun critique of U.S. gov't is unfair—But,

Since I believe that equatorial pulses of heated air hold the key to the Antarctic & Arctic weather stimulation and proper timing of the earliest springtime rains, as well as alleviation of the excess heat accumulating in the water around the South Pole Ice Cap we'll have to wait for the *partial realization* of the Qattara project to resolve this dilemma. Meanwhile we're not quite done criticizing (UNFAIRLY PERHAPS) the government that I love and admire. These next two criticisms are hardly even fair. They are in fact so unfair that I won't even bother to lay out a compensating list of things that we have done right—to balance the scales—lately. These were not even attributed to government officials at the time of their raising!

All that said, the U.S. position as pre-eminent superpower and counselor of note in the areas of water projects, led to the quiet shelving of three proposals to dam northward flowing Siberian Rivers to augment Russian agriculture. Although one apparently did go forward at a later date. The disquieting reason for halted planning was the fear of melting the Arctic Ice Cap and irrevocably altering the world's weather and climate. The supreme irony is that the melting appears to be taking place anyway—with (I suspect) little or no agricultural benefits accruing to Russian agriculture.

The other project that was stopped with a quiet word to the Saudi's, needs a bit more reasoning applied. The proposal which I remember only vaguely (and perhaps incorrectly) was to bring Arctic ice into the Mediterranean or up the Gulf or Red Sea to augment the Saudi's infant (and expensive) desalinization system. That proposal wavered and then fell by the wayside apparently lacking significant authority (meaning American) approval.

The extensive studies of the Mediterranean rainfall apparently lack for nothing in their extent, but they fail to account for the history and causes of the late season rainfall in the Mediterranean and the corollary dearth of early season rainfall. I'm referring to those causes which could make the late season rainfall later & more torrential and also to those causes which could make early season rainfall to be nonexistent.

To give historical substance to these kinds of questions—I quote from the Book of James, the 59th book of the Bible—James ch. 5 verse 7 "Therefore be patient, brethren, until the coming of the Lord. See how the farmer waits for the precious fruit of the earth, waiting patiently for it until it receives the Early and the Latter rain."

Single-hulled tankers don't have to be declared obsolete!

The doubts raised about the Saudis' tentative proposal went something to the effect that the cooling effect on the Mediterranean could make the scarce Mediterranean rainfall even scarcer & later. And those fears apparently came true with the increased melt water coming from the North the melting Arctic and Anemic Gulf stream both having their deleterious effect on the drought and flood spasms of the Mediterranean littoral. My position on the sea side of the Mediterranean debate is simply this: Proper temporary timing of ANTARCTIC_(NOT ARCTIC) melt water or ICE (remember those fear-inspiring freshwater lakes (lake Vostok) under the Antarctic ice, the ones that threatened to raise the world's oceans by 20 (?) feet or more overnight. An **empty** oil carrying tanker could refill **with clean Antarctic meltwater from** near the coast of Antarctica before rinsing the hold with seawater and further clogging of the world's oceans. That load of mostly fresh water **could** be discharged economically and effectively into the Afar Depression on the return trip to the Persian Gulf. There Mother Nature and equatorial heat could digest and assimilate, (with a little help from oil remediation experts). the tanker residue that continues to afflict Ocean phyto-plankton. It could be controlled in this Afar Depression as well as the Qattara Depression and potentially even Death Valley and the Dead Sea. All these water and oil residues deposited in the depressed areas of the world could be coupled with selectively timed micronutrient additions to /the Atlantic outside Gibraltar/ just properly before the Atlantic contributes to evaporative systems of the Mediterranean. The micronutrients could supply earlier solar gain to the evaporation from the Mediterranean but other coaxing would need to be supplied to augment the earlier timing of the rainfall cycle—more to this point later.

Needless to say the Saudi's with their 1300 + center pivot systems are already an asset in the effort to improve the timing and amount of the rainfall cycle. And the plans (perhaps preplans by Jordan and Israel is the preferred statement) for the RED SEA DEAD SEA CANAL would also help to improve the timing of the rainfall cycle but they are not enough.

Reraining documented with a chart of 100 years of Nebr. Rainfall!

The enclosed chart of 100 years of rainfall for Nebraska (below) is one of the clearer outlines of the benefits that can accrue with the application of the infant science of RERAINING. The first fifty or sixty years of the chart show the expected pattern of an approximate 20 to 22 year boom and bust cycle of rainfall.

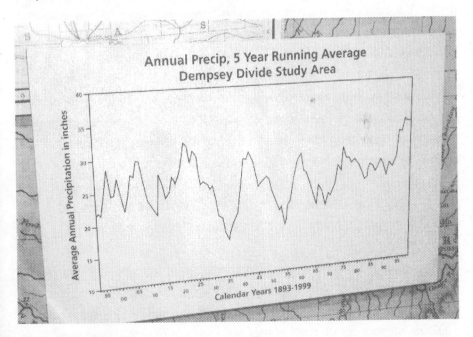

The last forty years are altered and increased by the phenomenal growth of the center pivot irrigation system and that this occurred in the presence of a world wide equatorial oil glut is a miraculous tribute to the (Science—Art) of Reraining & water mining. These later years show little sign of the expected minima and an overall increase in the average rainfall. I obtained it from one of the newest National Parks in Beatrice Nebraska. The Homestead National Parks tribute to individuals stands in stark contrast to the sometimes questionable achievements of governing bureaucracies. Altogether there is little doubt that Nebraska is one of the best governed states in the union.

Reraining commended for Southern California

There is one truly spectacular site for the explosion of the science (ART!?) of reraining onto the world stage. It qualifies under the ethical heading of "remedial" in more than two ways. You'll remember that my ethics does not permit unlimited "messing" with either 'weather' or 'climate'. We are only permitted to fix what we "broke". That site is Southern California specifically, The zone from San Diego northward along the coast to north of Los Angeles. The painful realities of this change will become obvious later in the script. The man-caused brokenness of the Sahara went mostly unremarked until about 50 years ago and its temporary expansion during parts of the last 50 years awakened alarm bells in many parts of the world community

Absence of reraining documented for television audiences.

The visible dramatic reverse proof of the effectiveness of reraining was shown by television media. A tributary of the Amazon was shown where the trees had been clear cut and we saw a dry riverbed with 1,000's of pounds of rotting fish lying in the naked streambed. If we still require proof of the need for the science of reraining it appeared in today's newspaper.

$800 million devoted to remedial reraining—Florida

The Kansas City Star Wednesday May 7th, 2008, page 2 EVERGLADES-reviving a treasure. "More than a century after the first homes and farms took shape in the Everglades, decades of flood-control projects have left the region parched and near ecological collapse. Now crews are building what will be the world's largest above ground manmade reservoir to restore some natural water flow to the wetlands.

Engineers 'built this thing beautifully,' said Terrence Salt of the Dept of Interior, referring to the flood control systems that practically drained the swamp to make way for development decades ago. 'But as we look back at it through the lens of our current 21st century values and understanding, . . . which leads to our restoration efforts now.'

"Water once flowed practically unhindered from the Everglades headwaters South of Orlando all the way into Florida Bay at the state's southern tip. But now when a hard rain falls, canals direct the overflow into the ocean to keep from inundating 5 million people who have settled in the area.

"That's where the massive reservoir just south of Lake Okeechobee will come in. It will store up to 62 billion gallons of water that would normally be channeled out to sea and instead divert it into the Everglades at various times to mimic a more natural flow" And after you store it for a period will the critical issue of timing be irretrievably lost in the same bureaucracy who will probably ignore the great contribution of the Gulf Stream to the other side of the Atlantic and is this a mega bureaucratic shuffle?

Potential Global benefits!

The missing piece of the puzzle is this. The murky organic contribution to the Gulf Stream determines whether it remains hot and on the surface thereby contributing to Continental rainfall and snowfall; or submerged beneath melting Arctic water. In this particular case James Hanson is correct in his expressed concern over the tipping point.

The timing of Florida release of organic material will inevitably determine the timing of the fall of precipitation on the European Continent and Russia. The reason that this concern is so pivotal is the necessary interment of Arctic Greens. The projected melting of the Arctic can be stopped almost single handedly by the careful conservation of Gulf Stream heat as it crosses the Atlantic to contribute to that layer of snow & ice that will prevent the climate insult of released methane to the global warming scene.

The three column article failed to mention either the Gulf Stream flowing around the tip of Florida, nor the recent anemia of that same stream. But warm organic laden water plays a huge part in an albedo colored view of the world's climate which is what this book is about. Nevertheless, 800 million dollars is a sizable commitment to the science of reraining.

Condensed summary: temporary ocean sealing contributes to semi-permanent sun-locking of land.

To sum up: there is an apparent consensus of scholar's and readers that the 5,000 year old insult to the Sahara is man-caused. There is a constellation of recent additions to that insult which do not have a full consensus—yet. They are Ocean sealing from 1900 to 2008 with its twin dramatic apices. One 1922-1936 and the other between 1978 and 1983 and culminating in the dramatic 500 year flood of the Mississippi & Missouri Rivers in 1993. SWAMPLAND DRAINAGE; which has been noted in many more areas than the Everglades, but is now giving way to an infant science which I shall call reraining. The positive example is shown in the 100 + years of Nebraska rainfall; the negative example is shown in the Amazon's basin development. The soon to be erected $800 million 62 billion gallon reservoir in Florida will be the bureaucracies shining moment of (corrected reraining or another example of criminal mismanagement and delay.) When I say delay I do not refer to strikes by the labor pool but to the larger question of the timing and strength of the Gulf Stream and of its European and Mediterranean influences. The silent corollary (or partner) to ocean sealing, increased radiative access to the lower levels of the atmosphere and even to below sea level has been siphoned off into concern about CFC's and Antarctica. I don't feel that I have dealt with it adequately but a 40 year record of the increase of skin cancer might help the consensus to form. Another corollary is personal—This is an almost insignificant addition in this list. I wasn't going to include it because I felt it was uniquely my own in 1980. (there isn't much chance for consensus there) that is the "milk carton experiment which confirmed (to me) "used oil" as a significant contributor to ocean sealing. But this mornings' paper KC Star 05/12/2008 pg. 2 Titled "Study Details Declines". In the midst of certain chemical toxic reductions there is this, "some problems highlighted include oil compounds from motor vehicles and shipping, which continue to flow into the waters."

This probably makes a consensus but the prolonged disability of a S.W. Bell head research Scientist during (70's-80's) derailed an effort to build a ceramic engine as a (substitute for metal) that is almost a consensus of intent.

To further amplify this concern is to note the 800 million India Indians that will soon have access to the motor car and you have ample grounds to search for a new consensus. About used motor oil—some of that concern may be alleviated by the recent recycling of old oil?

The prime corrective location for all of these sins (presumed and real) is the Qattara depression and I will discuss several methods of water resupply to the hottest most desolate spot on earth—just a few miles west of the teeming metropolis of Cairo, Egypt. This should be the largest climate restoration project in human history and significant parts have already begun.

I think I have laid out enough fears and negatives for now. I must lay out a semi-structured defense of my people and even of my government since my feeble plans can easily be quashed by a few significant throat clearings in Washington D.C. and in fact will require some forms of active approval from them.

Can Foggy-bottom (Wash.) see the forest and the trees?

First of all, TREES: Before Johnny Appleseed in Indiana, there were cultivated apple orchards in New York State. And before that, the Black Forest in Germany. Neither my cousin-in-law's 700,000 trees nor my Father's 1,000 or even the CCC's vastly larger and more systematic tree planting hold patents on respect for trees which probably goes further back than Sherwood Forest. It may even go back to 4200B.C. to those repentant souls who fled the encroaching sand of the newly denuded Sahara.

But the beginnings of a successful attack on the widening sands of the Sahara, were laid by some people who planted 17,000,000 trees in Nigeria and made some provision for tending them. A further continuation was made with the 30,000,000 trees planted in Kenya and there we have a Nobel Prize winning face to admire, and she has the grit to announce that a billion trees is the goal for 2008. Further the fate of Mount Kilimanjaro has been authoritatively removed from the hysterical side of the Global Warming debate, by the announcement that the disappearance of trees on its flanks contributed somewhat to a more than 100 year warming trend that accelerated the melting of ages-old snow and ice. But my addition to the debate—i.e. ocean sealing—may have contributed to the tribal desperation that resulted in fuel gatherers going further upslope on Mt Kilimanjaro, to secure wood.

But someone has to have credit for the awakening tree consciousness in Africa and I chose George H.W. Bush Sr. in case he helped make the CIA's contribution to "THE WEATHER CONSPIRACY_THE COMING OF THE NEW ICE AGE." This may ignore the work of 50,000 missionaries including my brother who died and was buried there— Zambia, Africa in 1995. That book woke me up—but even if Bush Sr. gets no credit for that—he gets my vote for his official recognition of the potential for oil spill damage in the Exxon Valdez disaster. Late as this awakening was, it was accompanied by a call for double-hulled tankers.

Why we may not need to scrap the antiquated single-hulls!

I don't know the target date or the completion date for this overhaul of the fleet of tankers *but I have a strongly repeated suggestion for the natural extension of the useful life of the single-hulled tankers. Use them to haul water from those threatening under ice lakes in Antarctica and deposit them into the soon to be dug canal into the Qattara Depression (or Afar) ever mindful that their timely flash of coolness could trigger a much wider harvest of the abundance of moisture which even now passes unhindered over the Sahara. It is probably impertinent to suggest this but the recent obscene surge in oil profits may be needed to finance the upgraded fleet. To prevent the premature retirement of the old one; If the existing single-hulled tanker fleet were enlisted to siphon dangerously growing pools of unstable liquid water from the piles of Antarctic ice and carry it first to the Afar Depression and ultimately to the Qattara depression before rinsing this time with clean desalinated water: then the useful life of these rust buckets could be extended and the ocean would be granted time to purge itself of these hindrances to evaporation. I hope this suggestion helps to alleviate skyrocketing costs of gasoline. The cautionary note that should be applied here is this: At what point does the application of Sprengel's law require parts of the oil residue (i.e. rusty metal) to supply the scarcest element for proper phytoplankton growth?!!*

There are a great many more proposals for the reawakening of water consciousness in Africa, especially the Sahel, and they include the headwaters of the Nile, as well as Central Africa. We'll mention some of them but only with the tacit understanding that any improvement in Africa's water resources will be a net gain for the whole planet—via its timely contribution to global air circulation patterns.

I can't begin the specific phase of this book (with your full attention and respect) with only the nebulous contribution of "global air circulation patterns," to convince you of your enlightened self-interest. The real "heavy" in this little drama is water. And James Hansen's "tipping point" is to be invoked to best describe my intentions for that new and ominously heated water around the Antarctic. You'll perhaps recall my semi-rhetorical reference to an alternate power source for movement of water (as opposed to trade winds) I.e. gravity.

Atlantic evapo-transpiration compared to the Dead Sea and the Atlantic of 1914.

Careful readers may remember that I laid out a so far unproven statement from the 1914 completion date of the Panama Canal—"If the Panama Canal was at sea level there would be a four-foot waterfall from the Pacific to the Atlantic Ocean." The enormity of the contrast may be highlighted by a short exchange with the Yale professor of Oceanography. Two years ago he was speaking about his concerns that Arctic melt water was piling up against the Panama Canal and the western side of the Atlantic Ocean.

> The decline of the Gulf Stream as noted could have been better described as the crisis of the absence of Atlantic evapo-transpiration. If you place that small four foot differential beside the estimated 10 to 22 feet of evaporation from the Dead Sea, the build up of heat in the Antarctic Ocean becomes a logical consequence of the temporarily stymied evaporation cycle in the Atlantic. We are going to re-establish that cycle both in the air and in the ocean, actively draining the Mediterranean into the Qattara Depression. We are going to encourage timely surface heating of the Ocean with selected micronutrients to make up for the missing organic green as well as the partly toxic effluent that comes off of America's heartland and perhaps some makeup for the belated renewal of organic green coming out of the Everglades.

Chapter 10

Hoped for effects and the science that could authorize it!

ut the completion of the deal will occur only when the evaporation and transpiration from the Qattara Depression reactivates the global winds both South Pole and North Pole. Obviously there is no proven direction and timing for the rising currents of air from the maximum solar converting area at or near the equator; but that knowledge will accrue as the first steps are taken. My own curiosity will not be satisfied until the compiled history of the circumpolar upwelling around Antarctica has gone on long enough to reflect some changes in surface evaporation hence a slowing in subsequent renewal in the rate of nutrient upwelling. Another way to examine the problem of the Atlantic evapo-transporation is offered by the evidence of 60 cm pileup of Pacific water near Indonesia at the start of an El Nino. If we had a satellite permanently tasked to examine the SOUTH Atlantic in the same way as the Pacific before El Nino we could have adequate grounds for the moral authority needed to rehab Atlantic evapo-transportation. All of these concerns reflect a legitimate concern for the nutrient health of the whole ocean.

For those who are rightfully confused about this paragraph. Schatzing describes a 1,000 year circuit of the globe by ocean currents. My concern is for that span of time that elapses between the salt laden submergence at an excessive evaporation point and a return pulse of upwelling at the poles of enriched material.

When I quizzed the famous professor for some reason that the Pacific might be higher—he gave me a lightweight answer . . . "it rains more in the Pacific . . ." True enough but not important to illuminate our search for the solution to rising oceans. The Pacific gets more rain of course because it is larger, but importantly for at least one quadrant—the southeast Pacific where it bumps against the 22,000 foot peaks of the Andes—hardly any transpiration time elapses between rainfall and its return to the Pacific. Additionally, the majority of Pacific typhoons expend their energy on the Ocean without appreciable gain to the lands around the Pacific. Also, we have already discussed the Atlantic advantage. That advantage once was the phenomenal amount of organic green contributed to the Ocean by the Amazon, The Congo, the Mississippi-Missouri, the Nile, The Niger and all the lesser rivers of Europe. Coupled with the unimpeded sweep of polar air from both North and South this weather machine once attracted waters from all points of the globe—but lately—residual polar (South) heat has been building to the hurricane point because of anemic equatorial evaporation. Now to get specific!!

I hereby nominate the "Rain harvesting Indian princess" who shall remain nameless but continue to be gainfully employed among the various medical and bureaucratic subdivisions of modern India. *This princess position is for the sole purpose of underscoring then eliminating the barrier that hot dry land imposes on the shoreward advance of moisture harvested from the sea's upper layers which have been murkily enriched by phytoplankton (or erosion) to helpfully magnify normal solar input.*

Mini-farming in the absence of regular rainfall.

Now that we've had our little fun-let's attempt some hard nosed common sense. My daughter with her husband and their five children recently returned to Southern Missouri-purchasing a mini-farm in the process of transforming themselves into hill-billies with a mild streak of survivalism thrown in. They departed from a four acre sandpile just west of Tucson where they had with patience and hard work raised chickens and vines with real melons and cucumbers and even produced a patch of green grass. On arrival (MO) they were welcomed by their neighbors and sought growing advice in various places. An older man who directed what might be labeled a commune sought to offer them helpful advice . . ."We have two rainy seasons here, spring and fall but you can't make a crop without irrigation supplementation anymore because the midsummer drought is too long." A personal interjection is in order here to underscore the point. A long-time friend of mine, a Mo. Beekeeper and past president of the state association, nearly quit the bee business after 2007 when he lost half of his 300 hives. After some deliberation he attributed the losses to an extended summertime rainless period. He has since rebuilt.

I attempted to raise tomatoes in my back yard last year after almost a 30 year hiatus while I raised children. I don't remember any water bills from those summers but I do remember eating fresh tomatoes in December of that year. Last year I watered extensively and harvested one tomato. The poor fellow looked lonesome sitting on the table-but when I looked at the $40 jump in my water bill-I didn't have the heart to eat him.

I've lived in the same house for 39 years-it's paid for. The real estate lady carefully explained to my naïve self that it was built with a "floating pad,"—which I took to mean that the sidewalls and the floor were not as solidly fastened together as they might be with another technique. Over the years I noted that on occasions that the rain exceeded 9 inches in a day; I would then have water in my basement.

Mini-fix-it attempts via Kiwanis.

I first noted this phenomenon about Sept. of 1981 or was it 1977-I guess that's just so much water under the (stairs). It became such a regular occurrence that my long-suffering wife refused to change the basement carpet until I got the water problem "fixed." I didn't want to fix the problem by hurrying the water down to the sea-there is too much of that already and I couldn't use my basement as a reservoir-not and stay wed, so I prepared myself to listen to the "experts". My Kiwanis Club (World-wide organization of thoughtful people dedicated to improving the welfare of children "One child and one community at a time.)" provided the experts from the MARC Mid America Regional Planning Commission. They brought a colorful brochure including seeds and talked about a "Rain Garden." Essentially they wanted the citizens of Kansas City to place 4-50 gal. Drums at each corner of the house to trap the rain water as it came from the roof and down the drain. All the valves were to be supplied-including natural mosquito shielding. I was so sure that it would help that I went to the local hardware store where they told me the units could be purchased

Maxi-fix-it, ridiculed in 1996.

The Local hardware store hadn't heard of this plan or I might not be telling this story. But back in the Kiwanis Club and the experts, They called their plan "Carbon Sequestration". During the question and answer session, I picked up on the "Carbon Sequestration theme and reminded them of the Magnificent Carbon Sequestration plan that was placed on the political table during Bob Dole's failed presidential bid of 1996. Essentially the plan had called for a canal to be built from the Missouri River Impoundment in South Dakota across Nebraska and Kansas to provide much desired relief to those perennially water short but nevertheless productive soils. The Kansas City Star ridiculed the idea as Bob Dole's personal Pork Barrel project and the idea died along with his election bid.

Having grown up in that area I'm well aware that there is only enough natural rainfall to spur vegetation growth for about three months. That monstrous high plains area with its shortened growing season happens to coincide with a dramatic annual springtime drop in CO2 concentration-from The Coming Storm, Mark Maslin, p. 93. I believe that drop is no accident and could be extended here as inexpensively as anywhere in the world. It could become a prime Carbon Sequestration area with a minimum of invested capital. The objection that was raised under the cloak of logical opposition noted that the Missouri River Barge Season might be shortened by this diversion. When you note that the water is not removed from the watershed, but is instead placed where it could only reenter the watershed with several potential re-rainings the project should have received the unqualified endorsement of thousands.

Mark Maslin and the hidden mini-truth of the Co2 charts.

It's now time to soberly examine Mark Maslin both pages 92 and 93 because I believe this material will hold the clue to halting the seemingly inexorable climb of the ppmv (of CO2) and even of potentially reversing *it even in the face of continued expansion of fossil fuel usage here he is. "Should we be worried about global warming? Yes, and the reason why we should is that there is clear proof that atmospheric carbon dioxide levels have been rising since the beginning of the Industrial Revolution. The first measurements of carbon dioxide concentrations of the atmosphere started in 1958, at an altitude of about 13,120 feet (4000 m) on the peak of Mauna Loa in Hawaii. The measurements were made here because the area is remote from local sources of pollution. What they have clearly shown is that atmospheric concentrations of carbon dioxide have increased every year since 1958." (Twin graphs will be inserted below so that you can decide for yourself, but this material is not quite true. There is an absolute pause in Mark Maslin's graph of CO2 concentration in the atmosphere: Mauna Loa curve. This graph shows an absolute drop in roughly the year 64 to 66.*

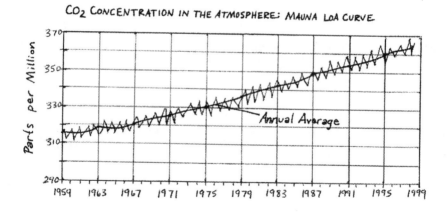

drawn from <u>The Coming Storm</u>, by Mark Maslin

Amber Dleathing

The second graph labeled global atmospheric concentrations of CO2 and described as records from ice cores show that levels of global atmospheric carbon dioxide have been increasing over the last hundred 20 years. What I see in the graph is a perceptible uptick in the amount of CO2 captured in the ice.

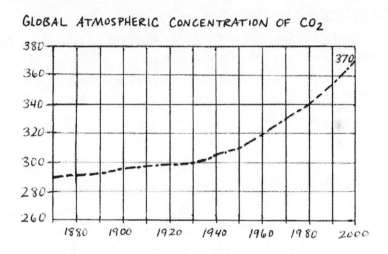

GLOBAL ATMOSPHERIC CONCENTRATION OF CO_2

drawn from <u>The Coming Storm</u> , by Mark Maslin

Amber D Leathers

This occurred roughly beginning in 1965. My account of a dramatic shift from particulate emissions beginning in the early 1960s highlights and explains the absolute drop in airborne measurements of CO2. But **I'm not advocating the return of above ground testing of nuclear bombs nor the building of a second Interstate Highway System or even the pre-fifties method of farming as some kind of panacea.t** We are only highlighting the greater question which lies behind the cause of the drop in CO_2 airborne. The trigger for the hyper activity of the ocean surface could have been any one of several kinds of particulate fallout . . . Construction dust, nuclear fallout, and coal fired power plant emissions are three of the more prominent candidates. Even the belated oceanic digestion of tanker residue is fair game in this appropriately Sherlockian search for Prime Causes Let's go on briefly with Mark Maslin." The mean concentration of approximately 316 ppm by volume (ppmv) in 1958 rose to approximately 369 ppm peak in 1998 the annual variation (drop) seen in the graph is due to carbon dioxide uptake by growing plants. The uptake is highest in the northern hemisphere springtime—hence every spring, there is a drop in atmospheric carbon dioxide, . . ." this material and atmospheric graph doubly underscores the utility of Bob Dole's carbon sequestration plan for the Missouri River.

It only takes water on the high plains to elongate and expand the growing season for that vast expanse of territory. We'll approach the water problem from the point of view of Southern California with a triple threat proposal—later—but for now. We've strayed from the rain barrels, but not from the purpose of the rain barrels. When I could not find them in the store ready-to-use, I stopped to think. After a bit of math, admittedly rough and rusty, I concluded that my 50 gallon drums would only delay the first inch of rainfall leaving my roof. Assuming that a healthy lawn could only absorb 2, 3 or at the most 4 inches of rainfall; that left as much as 5 inches to contribute to the epidemic of flash flooding, as well as the 8 inches of surplus rain leaving my roof, not to mention my sidewalk and driveway. Consequently, I quit searching for 50 gallon drums.

Cisterns grow up to meet the size of the task

On the way to bigger thoughts I heard a report coming from Australia, the maximum flood and drought sub-continent capitol of the world. Houses in drought prone areas there are attaching 260 gallon retention basins to their homes. Now that's real "rain harvesting." Very applicable to the

Southern California multi-million dollar homes regularly threatened by the Santa Ana winds and fires. When you link this idea to the promotion of Death Valley as an evapo-transpiration HQ that should be enough hint that government then would inaugurate the building of the '3rd or 4th largest river', at once. Additionally, the government would probably jump in with all four feet and mandate 1,000 gallon rain vats at the corners of Southern California housing. This idea will receive its proper conceptual foundation in a separate section dealing with the arid Southwest.

Meet the cistern's Big Brother

One more idea on the subject of rainfall retention and utilization. This one is not mine because it apparently originated in or near Washington D.C.—Foggy Bottom USA. The basic premise behind the idea was that a house could be heated or cooled by a figure 8 swimming pool with limited or no access from one side of the pool to the other.

One end of the pool-the sunny side would have reflector augmentation to store the summer sunshine in the depths of the blackened pool. This heated reservoir—water-sealed somehow to minimize evaporation losses-would be harnessed to a heat exchanger to heat the house in winter. The other end would have a thermostatically operated spray nozzle designated to place a fine mist into the air at those times when the air temperature was low enough to transform it into snow, sleet, or hail and then to fall into the basin to be stored for summertime cooling needs. The proposed efficiency of the system would be heightened by the capacity of the ice to be stored above the confining sides of the pool. Additional unquantifiable assets, (efficiencies) would arise from the removal of summertime heat excesses into the pool and the release of the heat of condensation into the wintertime air thus ameliorating both sets of extremes. Remember Hawaii's climate? Or better yet Jamaica's modest extremes? I'll tell you about that in just a bit. I'm not trying to create Utopia here-I'm only ethically licensed to redress the *temperature extremes brought on by mankind, tree-cutting, oil spills, hardened surfaces with inadequate flood-impoundment reservoirs.*

When I glommed onto the idea I multiplied it times the number of homes in the Metro Area which is much larger than the 323 square miles allotted to KCMO, and tried to conceptualize the effect of 300,000 homes spraying a fine mist into the air as the first blast of winter hurtled down the Missouri and spilled out over Kansas City proper. Bearing in mind the 78 degree below zero chill factor that I experienced so indelibly in 1978-it seemed a truly marvelous idea with nearly a zero handicap in the yet-to-be-evaluated particulate-emission debate. That is only wintertime. The summertime effects could be much greater depending on how much "Rain Harvesting," could be built into each system. **No one could even guess at the size of the reduction in fossil fuel use for a city of Kansas City's size, but a gleam of insight might be derived by a per capita comparison of heating and cooling costs for a coastal city such as San Francisco Jamaica has modest temperature extremes.**

I already mentioned amelioration in relation to Hawaii. Let me share a personal story from Jamaica in the early 90's. Our church group went in early February to help build a doctor and dental clinic for poverty-stricken Jamaicans. We **mostly** *worked mornings and toured afternoons with the evenings reserved for church. The services were entertaining especially the ritual of the offering. The conversations at the services and the next morning in the work routine (which we shared with the Jamaican's), mainly focused on their desire to come* **to America**. *For some mostly unknown reason I was asked to address one of those evening sessions. Thinking to deal with the not-quite rational desire to come to America-I had already warmed to their mild February weather-(this was before I had lain wide wake till 3 A.M. to the tunelessness of their multiple boom-boxes; they ebbed a bit at that hour only to be replaced by the crowing of roosters]. I'm sorry now if I appeared to be trying to stem the tide of immigration, but our dialogue ran something like this.*

I asked the audience of 50 or 60, "How hot is hot-weatherwise here in Jamaica?" There was a spirited response, finally resolving itself to 85 degrees, with a small minority holding out for 90 degrees F. I *then asked.* *"How cold is cold?* There was a bit of thoughtful silence with a sprinkling of tentative answers-split between 60 and 65 degrees. Finally an older lady in back spoke with authority. "55 degrees!" When I suggested the working range of temperatures that I had experienced in Kansas City, Missouri-from a chill factor of 78 below zero F. to the six week extreme of 100 degrees + there was a disappointed silence. We Northerners had already worked through a mid-70's shower that had sidelined them. I'm sure that 300,000 Rain harvesting figure 8 swimming pools-(air conditioning and heating combined units) could not achieve the welcome ease of the Jamaican climate; but it could subtract from the continental extremes that contribute to the Globe's weather extremes.

Chapter 11

GLOBAL WARMING BOOK III CARBON SEQUESTRATION BEFORE IT IS SPILLED OR SPENT.

top. Stop. Stop. Stop!! I could write this word a million more times and you would probably not yet grasp the message. But I think I may have found (one of) the *"Elephant's in the Kitchen."* I hate this next section because I'm apparently attacking my best friends. (The governing infrastructure of the greatest nation on earth) when they (possibly including you Mr. Gore) turn a deaf ear to my apparent attack-who then will listen to the message?

I'll begin with an obvious non-sequitur. Did you know that Mercedes Benz established the all-time high in vehicle miles per gallon (either 3,000 or 5,000 mpg—I forget) established on a closed circuit oval track by a vehicle that never stopped? I believe it accelerated from 5 mph to 55 mph and shut off the motor and coasted until it returned to the 5 mph whereupon it restarted and began the cycle again.

Did you know that my home state of Nebraska with an area of almost 80,000 square miles has more than 80,000 stop signs? That is a guess *since I obviously didn't count them-but I know that much of the state is laid out in mile square grids because there are fewer natural obstacles to road-building than in almost any other state. This means that much of the state's traffic is subject to the dreaded prospect of a potential T-Bone crash, at speed, with often fatal consequences.*

In Nebraska its STOP or else but Mercedes suggests 1,000% improvement is possible

I remember a time before stop signs proliferated in the state of Nebraska. That again is a guess. But in the early 50's one old highway from Friend to Exeter crossed a newer more traveled route proceeding south to the town of Milligan. A friend of mine proceeded west on the old highway and because of the burgeoning corn fields couldn't see the approaching car at right angles to him. After the crash I visited the irrepressible young man in the hospital-he fully recovered. There was no stop sign on any portion of that crossing, but shortly after the incident stop signs began to sprout at almost every intersection in the state.

Cause and effect magnified from local to national.

The incident proves that the legislature was very aware and doing its job? Right? Think again-this time with a broader perception of cause and effect. Not long after this-Ford came out with the '56 Ford-seatbelts and dashboard padding-good idea! Not so fast-Chrysler came out with the then ultimate muscle car-the Chrysler 300-and captured the hearts of impressionable young men who were-guess what-tired of sitting behind a stop sign-more often then not with dust irritating their lungs and ethyl gas vapors with lead potentially influencing their central nervous systems.

There were other (local) responses to T-Bones (crashes) including a wholesale community effort to go out to the offending intersections and clean them out with corn knives. The most telling response was to leave the source of danger and seek safer venue-city streets. But the complex dance between governing authorities and builders of automobiles that resulted in a plethora of muscle cars became the dominant response.

I can't in good conscience expect John Q. Citizen to grasp this argument because I had 35 years of cross-dock traffic experience on the Yellow Freight Dock and still couldn't conceive of a better alternative until I retired and traveled to Norway.

Cairo and Norway—a different idea.

But since I am about to sprain your brain with Norway, let's take a preliminary glimpse at Cairo, "Egypt. My wife and I arrived there in the spring of 2002. We were told by our tour guide that we were the first Americans to enter the country since 9 11. Armed vehicles accompanied our bus as we crossed the Sinai Desert. I tell you that so that you know I was not the least bit sleepy or inattentive as we pulled into the city limits during rush hour. Cairo is a city of 15,000,000 confined in approximately 250 square miles. Kansas City by comparison is less than 500,000 in 323 square miles. Cairo has probably 1/3 the traffic lights of Kansas City and less than half of those traffic lights were working that afternoon-and all of them—red, green, yellow, or black-were ignored by traffic that proceeded at a stop and go snail's pace. Every auto but one had at least one dent and many had three or more-none deeper than three inches that I could see. A grand Rolls Royce (oversize) had the deepest dents and one car—a brand new Mercedes Benz had no dents. Six lanes of traffic merged into one in a length scarcely longer than a city block. Now that you know that there are other systems. (??) I'm ready to take you to Norway.

Five of us arrived in Oslo, Airport. Everything normal-even luggage. We were met by more relatives-a family of 6 with one guest-total 12 persons with ample luggage. We boarded a 16 passenger Sprinter Van with ample room for two weeks of luggage. Since I had already heard of the vehicle-High mini-van mileage for a much larger vehicle-I let my curiosity show. The host promised to translate the liters and kilometers into miles per gallon for my American inquiry.

After three days to get acquainted with the estate including mini greenhouses-beautiful Drammen River flowing by at the foot of the lawn and multitudinous trains that quietly hissed by within 30 yards of the house. It rained every day and every day we went to the garden and picked that day's supply of fresh fruits and berries including grape-sized gooseberries that were sweet enough to munch right in the garden. The hiss of the trains was not the thunderous click-clack boom-characteristic of American railroads and became easy to get used to. It was a beautiful home setting for the beginning of a 4,500 kilometer trip up the rocky spine of Norway past the Arctic circle and across by way of ferry boat to the well-populated islands north of the Arctic Circle.

After driving all day apparently without filling up, my host finally filled up and courteously transferred the results into English. His result: 33.6 miles per gallon. I already had an American figure for a similar or smaller configuration of the same van-approximately 20 miles per gallon with a similar load. We had 12 people and their entire luggage plus enough food for the week and three bundles of parasailing equipment. What could be the reason for the apparent 50+% increase in mpg?

Here's what I found! Almost no stop signs. Lots of roundabouts and very few traffic lights. Oh! Instead of climbing a mountain at every opportunity the Norwegians dug a tunnel and reduced the demands of gravity wherever possible. Our host-driver was a tunnel-washer by trade and the family showed scrupulous awareness of the necessity of reducing the level of befouled air by closing the vents at the entrance of every tunnel and opening them at the exit.

The curious feature of the few traffic lights that we experienced—No one seemed to feel the need to do a jack rabbit start after the light-almost as if they were aware of the cloud of mood-altering pollutants that they would have left behind for their fellow-traveler. On the other hand there was plenty of room for aggressive driving. The bridge over the Drammen River, a block from their home, was a quarter mile long and one car in

177

width. Their cul-de-sac joined the main road through town less than 20 feet from a blind overpass for the train. In America there might have been a stop sign or even two, or maybe a red light with a **no turn on red** sign. On this road there was nothing-(not even a yield sign) but it did join in such a way that the merger was executed at an acute angle . . **That's it-now I can see it all over Norway-they just don't engineer their roads at 90 degrees of opposition—certainly a lot less of an opportunity for a T-bone collision at cruising speed.**

There were plenty of roads that weren't wide enough for two vehicles to meet at speed. But there I saw common sense and resentment-free yielding by either vehicle. To fully capture the flavor of Norway you had to see the children of the family fanning out freely at the several stops we made as we traveled upstate. Three girls, 13, 11, and 7, and one boy aged 9 found ripe berries at nearly every stop and ate them unhesitatingly with no qualms about roadside debris or toxic defoliants. What I saw on that 4,500 Kilometer trip was the **Perfect picture of a state that saw no reason to 'go to War' with (unnecessarily irritate) its citizens.** The same picture came through quite clearly in church. The 11:00 service was evenly divided between men and women; and no man was distinguishably overweight.

Co2 crisis gets personal

Do we have enough here to lapse into my personal favorite-the mildly prescriptive mode? I think not yet. Before I do that I need to recapture the time when Yellow Freight launched the Teamster's Union into the Machine Age with all four feet. During a given period of time the percentage of forklift freight increased rapidly and the company belatedly discovered that they were idling their workers for unnecessarily long periods of time while 6 forklifts tried in vain to answer the calls for assistance of 30+ men. During a 6 week period of time the forklift count rose from 6 to 36 machines-eventually to reach as many as 60 in number. During that interval there were some temper outbursts that were totally normal and some that seemed to me to be purely chemical in origin-the logical consequence of being stalled in traffic behind the exhaust of one or more forklifts. We had our own mini-CO2 (global warming) crisis.

When we noticed and avoided prolonged idling stalls-the anger level subsided and a much-improved order was restored to the dock. By way of explanation: CO2 in the blood stream becomes carbonic acid-the one absolutely indispensable ingredient to enable the 7 foot+ high jumper-the fabled jam in basketball and the numerous spectacles on the football field. We all love it and need it-witness the overflowing crowds at football games-until we OD on it-with the consequent obesity and diabetes-but relatively few people suspect that we are being force-fed artificially at the standard All-American traffic light. The pre 1966 chemistry involving lead in the exhaust and a complex mix of CO2 and carbon monoxide produced an uneasy balance of toxins. The newer "Cleaner" exhausts just happens to propel us into the accelerating mode and hence we get road rage as our reward. There is a handy temporary solution-just **"stand back from the offending exhaust-particularly at an uphill traffic light** and you'll find the mood of your day stays at a more even keel.

CO2 goes Global and Prehistoric

A more adequate solution is on the drawing board in the form of the hydrogen burning engine-it may or may not show up. But there is a better solution for all concerned; **and trucks have had it for years. JUST RUN THE EXHAUST STRAIGHT UP!** I have been reluctant to propose it for years-out of sight, out of mind-is simply not adequate. But if we make a serious commitment to the deserts of the earth-we will then regard CO2 as an indispensable ingredient (fertilizer) much desired and in fact treasured. The answer to the next question from the reader who is two steps ahead of me-where does that straight up CO2 go? I have no answer-rather several illuminating but incomplete observations. I suspect that any answer offered now would be at least slightly different than the years that CO2 was exhausted with lead as its partner. But if we revisit the Mark Maslin charts from a few pages back we can safely conclude that some of that airborne CO_2 is safely confined in Greenland ice. Although, I know that CO_2 levels have been measured at the peak of a Hawaiian mountain and are said to be relatively the same worldwide. I also know that increased levels accelerate plant growth. Chris Horner, (no relation) in **Politically Incorrect Guide to Global Warming** p. 66 . . ."Plants appreciate warmer temperatures **(As well as higher CO_2 concentrations.)**" (From pre-history—Carboniferous) also courtesy of **"The once and Future Story of Ice Ages." P. 149** "North America and Western Europe . . . . cyclical: beds of coal alternate with marine sedimentary rocks such as limestone or shale in a pattern that is repeated many times over."—the CO_2 comes down out of the air and is fixed in plentiful quantities in marshes that become coal fields at (Slightly above or below?) sea level. But the scientific data of regional highs and lows as well as seasonal highs and lows (of CO_2 concentrations) hasn't yet achieved the public definition and clarity to support the ambitious leaps that I make-next paragraph!!

Sprengel's Law, the Dust Bowl and Giant Cabbages.

There were stories and pictures in **Popular Science** shortly after WW2 about cabbages that had to be carried in wheelbarrows. They were grown north of the Arctic Circle-also N. Scotland. At the time of first reading-my early teens-I was amazed and only slightly mollified by the explanation-24 hours a day of sunshine. But from **"Freaks of the Storm"**, Randy Cerveny, p. 239 "Edward Teele, a writer for **Popular Science Monthly** in 1935 wrote that, prior to the black blizzards of that year, a county road project near Hutchinson, Kansas had been tasked with the removal of 10,000 cubic yards of dirt. Before the project could begin in earnest, one of the black rollers had hit and, when the workers returned to their job, they found that all of the dirt they were to remove was already gone. The incredible windstorm had simply carried the dirt completely away." Where did it go? Like CO_2 nobody seems to know or *care. But if you can believe the story of the decimated Dust Bowl farm halfway up into Canada-its not impossible to believe that some of that dust contributed fertility even as far north as beyond the Arctic Circle. This truism says nothing about the contribution of dust to the surface of the ocean, nor to its ability to increase the alkalinity of the ocean. There-that specific plains dust had to contribute a measure of murkiness hence contributing to the-evaporation cycle. Its iron-rich nature probably catalyzed phytoplankton growth-aiding the transpiration cycle. All of this is problematic but we know for certain that it spurred the growth of global cooling theories that culminated in the scare scenario of "**Nuclear Winter.**"*

Concept of reraining transferred to North Africa

Part 4 Water to the Qattara Depression

The central remedial thesis of this book is that human events within the past five to ten thousand years have contributed to driving the jet stream "nature's rainmaker" north and consequently inducing the vast hot barrenness of the Sahara. Some portion of this conclusion is arguable in that some climatologists would argue that it was the disappearance of the vast ice cap on North America that drove the jet stream inexorably northward during the time when the Sahara was covered with grassland and trees. Nevertheless, I am proceeding with my revision of the 100 year old project. I am determined that it receive adequate conceptual ground work. Its success will be of greatly wider benefit than to its immediate neighbors.

My reference for the following is: Answers.com Wiki Answers Categories Science and Technology Environmental Issues What is the Qatar Depression Project?

"The Qatar (or Qattara) Depression Project was a proposal to build a hydroelectric power plant by connecting the depression (located near Alexandria in Egypt) to the Mediterranean Sea by an 80 km tunnel. Since the depression is over 130 M deep and the climate is very hot and dry, the water level will reach a steady state, balanced by evaporation.

A secondary benefit is that the evaporation will increase the rainfall and decrease the temperature in the area, both welcome changes. There are no permanent settlements in the depression, which is partially covered by hostile salt beds, so no population (besides a few nomads) needs to be moved. The project is similar to a better known plan of connecting the Mediterranean to the Dead Sea in Israel. Both proposals are nearly 100 years old."

All of the above is true and helpful but it is not nearly large enough or complete enough to remedy the globe's water crisis of increasing ocean temperature, nor will it give us enough choices to avert James Hansen's "tipping point." In today's climate the project must provide welcome relief to those nomads as well as enthusiastic applause from Cairo's multitudes. I believe these ideas will qualify.

For purposes of comparison and contrast let us review a lesser project: The Damming of Gibraltar. Online: Posted: Wednesday May 30, 2007 6:04 p.m. Post Subject: The Daming (sic) of Gibraltar.

Consider a silly theory for size and scope.

"Below dear reader you will find a straitline (sic?) {strategic?} proposal for one of the largest geo-engineering projects of all time, the daming (sic) of the Gibraltar straits to control the direction, volume, and makeup of the water exchange between the Med Sea and the Atlantic Ocean. The authors in dead earnest propose that this megaproject must be done to prevent the onset of the next ice age in 2090 due to global warming (and you thought my hype was out of bounds) and the Aswan High Dam in Egypt."

Their argument: "If the Mediterranean Sea continues to increase in salinity, shifting climatic patterns throughout the world may cause high-latitude areas in Canada to glaciate within the next century. The Mediterranean is starved of freshwater by human activities: most of the annual flow of the Nile River is now used for irrigation and no longer enters the Sea. Surface evaporation losses are also increasing as the surface warms due to rising CO_2 concentrations in the atmosphere. Consequently, the Mediterranean hydrologic deficit is steadily increasing. The deficit is the difference between the larger amount of water loss by evaporation and the smaller amount received from rainfall and river inputs. The difference is made up by a two-way exchange of water with the Atlantic at Gibraltar. Barring a significant change in regional atmospheric circulation, these two human modifications of the environment will cause the salinity of the Mediterranean to increase for some time as fossil fuels are consumed . . ."

Maybe hot exiting Med Saline water could trigger timely precipitation in the Arctic!

"The higher salinity will lead to a larger volume of the Mediterranean outflow at Gibraltar, which will modify high-latitude oceanic-atmospheric circulation and, in effect initiate new glaciation. This hypothesis arises from a recent study of climate conditions and inferred circulation changes that probably triggered the last glaciation {Johnson, 1997}. The hypothesis will be tested in coming decades. If it is validated by the onset of ice-sheet growth in Canada and cooling in northern Europe, a partial dam at the Strait of Gibraltar could be constructed to limit the outflow and reverse the climate deterioration, thus holding off the next ice age." I have to halt here! There is more to the quote (study?). The cavalier attitude gets me! There is no mention of effect upon shipping and no estimate of what this does to the rainfall pattern. And if it lowers the Mediterranean by how much???? *And what if that stoppage turned the entire Mediterranean into a Dead Sea or a stinking mass of raw sewage?*

Brief recap of unnatural '93 Mississippi Flood

If the most powerful man in the world-Republican-could not suggest to a Democrat to "fill the busses and run for high ground!" Nor could he after reflection volunteer 13 or more trainloads of Texas fill dirt to bring the 7 foot below sections of New Orleans to a safe habitable 7 foot or more above sea level. In the presence of an urgent need for a life-saving consensus there was almost no significant communication with New Orleans! Thus; what kind of consensus could be achieved among the perilously water-starved countries of the Mediterranean basin which MIGHT reduce their already stagnating water supply.

This irreverent question persists in my mind! Is there an obscene amount of ill-gotten gain to be had from repeated inundations of the city of New Orleans? Thus it strikes me that this dam project has less chance of success than a snowball in the Saharan sun. Particularly the part about seeking a motivating consensus from the vitally concerned nations.

But one observation is of some interest? The potential of hot saline water from the Med contributing to a wealth of arctic precipitation and hence a future threat of an 'Ice Age'. Didn't we have some rather large scale complaints about excess melt in the Arctic? We'll speak some cool calm words of guidance when the duel Canal/tunnel to and from the Qattara Depression is completed. The return tunnel for the hot dense saline material comes equipped with a shut-off valve which is to be governed by a committee of Northern Hemisphere countries which have the right to assess the effects of hot saline upon those most frigid climates. So far what is lacking for effective decision making by far North countries is the time lapse for the Med's sinking hot saline water and its North Polar reappearance. Norway, Sweden, Iceland, Canada, Greenland and Russia come easily to mind, but England, Germany, Scotland and Ireland are not far behind. ***And due to their vital interest in the height of the oceans, allow the Netherlands to chair the group.*** Let's continue with the quote.

"If a decision to build the dam is delayed, a more worrisome concern is the scarcity of petroleum supplies that will probably develop in the next century. This could lead to political or military conflicts that would increase the difficulty of organizing the broad international effort needed to carry out the project. The sooner the decision is made, the easier it will be to plan and complete the dam. The Aswan High Dam's effect on Mediterranean salinity will trigger glaciations much more quickly than would CO_2 warming alone, thus providing a compelling motive to build the remedial dam, and quickly, while petroleum supplies are still plentiful and society is stable."

Next, is his quotes-not mine. "I'll still say removing Aswan would be a lot easier/cheaper, and if that is not an option, diverting deep saline waters into the Quattarra (sic) Depression through a tunnel connected to a pipeline into the deep basin would act to help in three ways, it would remove a good volume of the deep saline water they are so concerned about thereby restoring the saline balance of the deep basin areas of the Med Sea. Given that the Quatarra (sic) depression hydroelectric project would produce substantial electricity it would even ameliorate the loss of the Aswan high dam and could make its removal or at least reduction in the volume of Lake Nassar a more economic alternative for Egypt. "(considering that the figure of 60% is generally offered as the amount of food lost before it reaches the table-for lack of refrigeration—electricity generation would quickly pay for its end of the project but the truly bountiful results would resound far beyond the borders-that's my prediction-hardly a proven fact.

I hesitated long before including this material which seemingly lightly dismisses the welfare and lives of 1/3 of the people of Egypt-(the amount of population growth attributed to the nation of Egypt because of the Aswan High Dam). I must remind you now that the ruling thesis for my life since high school days has been three words adopted from Albert Schweitzer's life in Gabon. "REVERENCE FOR LIFE". If that motto is found to be absent from this manuscript or any of its applications-then I will be adjudged to have gone seriously astray.

But because its premise (hot saline water exiting from the Med will produce a future gusher of wintertime precipitation in the Arctic) has been superseded by reality (extensive Arctic melting). From Science News May 10th, 2008 pg. 15: "It took almost a month for meltwater to accumulate atop Greenland's ice sheet in the summer of 2006. It took only 90 minutes for all that water-a lake so large it could fill New Orleans' Superdome more than 12 times over-to pour through a crack in the kilometer-thick ice below it and drain the lake dry.

At its height, the torrent exceeded that of Niagara Falls, and its rumbling triggered instruments nearby. GPS equipment indicated that the westward flow of ice in the region briefly surged, a sign that the water drained down to the bedrock and temporarily lubricated the boundary between ice and rock.

Some scientists have suggested that an increased number of similar events could spur a collapse of much of Greenland's island wide ice sheet, leading to sudden rises in sea level." . . . . We have all been subjected to the melancholy sight of 25,000 polar bears with scarcely enough ice floes to exercise their considerable hunting skills . . . (A wise extension of the concern for polar bears might have taken into account the various threats to ocean phytoplankton which forms the dietary base for the seals which are the mainstay of the polar bears diet). A brief rereading of Robert W. Felix's NOT BY FIRE, BUT BY ICE would tend to suggest a different source for extensive Canadian Arctic icing—but we've shot enough holes in the Gibraltar theory for now.

Suffice it for now to suggest that not enough is known about the timing of Arctic evaporation and subsequent precipitation (rain or snow), to prevent the next ice age via Atlantic causation; but I think we have suggested three very important factors which have in the past exhibited some (mostly unconscious) influence. They are: Hot saline Mediterranean exit material. Organically enriched Everglades outflow-progressively limited since the fifties and only now being partially restored. And third-the great Unknown-the Regulation (both hurrying {levees} and delay {dams} of the mighty Missouri-Mississippi. Coupled with the toxics-both medical and agricultural, the fertilizers, animal, chemical and human, and not least, the influence of oil, both used and freshly extracted. (Ixtoc 1 or a billion oil changes, take your pick) and don't forget 5,000 tankers regularly flushing their tanks at the end of the payload.

When we examine the Mississippi for its influence on albedo we have a shortage of useful data to hang our speculations upon. Here is a beginning insight: from page 2 of THE KANSAS CITY STAR, Wed. June 11, 2008. "SCIENTISTS BLAME FARMS FOR LARGE 'DEAD ZONE' THEY PREDICT THE BIGGEST-EVER OXYGEN-DEPRIVED WATER AREA IN Gulf of Mexico. By Sarah Lohman McClatchy Newspapers. Washington scientists predicting the biggest-ever 'dead zone' in the Gulf of Mexico are blaming Midwestern farms. The corn grown for ethanol creates high levels of pollution that escape down the Mississippi River, they said.

'In the past several years, there's been an expansion of corn, which has the highest fertilizer per acre . . . and that's for biofuels.' said R.Eugene Turner, a Louisiana State University professor who directed the study into the gulf's water quality.

Turner's research team reported this week that an area of oxygen-deprived water in the Gulf of Mexico is expected to grow to more than 10,000 square miles this year. The largest the area has ever been measured was in 2002 when it was about 8500 mi.^{2} . . . . The dead zone is caused by an abundance of nutrients from sources that include fertilizers needed for corn."

Proper interment= appropriate carbon sequestration

The incredibly useful question which needs to be directed to his fascinating description is this. Does the *'DEAD ZONE' DISAPPEAR OR MERELY DWINDLE SHARPLY WHEN THE MISSISSIPPI DEPOSITS A SEMI-NORMAL LOAD OF SPRING-FLOOD SEDIMENT?*

It may sound like whimsy but what is needed here is a proper burial. I'd like to refer you to the book called deserts by Peter Aleshire

Pub. 2008 pages 148 and following, "Sand dunes swallow dinosaurs. Despite the scale of the great Gobi Sand dunes, it is still startling to learn that more than 65 million years ago those sand dunes were big enough to bury a dinosaur in an instant. At least, that is the conclusion of an international team of scientists who published a study in geology . . . Roy Andrews Chapman, . . . first discovered the surprisingly well preserved dinosaur bones and eggs in the dry sands of the Gobi Desert. [scientists] found they could study tiny skeletal structures in the very bones including features smaller than the one of the letters in this sentence. Scientists have found the remains of fierce meat-eaters larger than Tyrannosaurus rex, perfectly preserved with scarcely a bone missing. They sought to explain how an entire dinosaur was killed and buried so quickly that not even the quickest scavengers or bone eating bacteria could get to the remains before the fossilization process started. The scientists proposed a massive sand slide . . . The overlying sand then protected the bones of the buried dinosaurs from scavengers, and the lack of oxygen deep beneath the sand prevented bacteria from breaking down the bones . . ." *That is only one side of the remedy for the multiple dead zones in the ocean. Another side will be found in the timely remedies for the global winds.* Now back to the dead zone study.

"The nutrients act as a fertilizer just like they would in a corn field,' causing excessive growth of algae on the ocean floor, Turner said. The {dying} algae consumes oxygen, forcing aquatic creatures like fish and shrimp to flee or die." (It's not the right time to bring up Sprengel's law—but it is well-**known** that wind acts as an aerator and this year's wind certainly qualifies on land—I guess it is okay to confess that I at least do not fully understand the eutrophication —{the long time conversion of a lake to a marsh by progressive enrichment particularly as to whether much death is involved} process or even whether it might apply here. " he said the increase in corn production as well as heavy rain in the Midwest this spring had caused a higher concentration of fertilizer in the Mississippi River. Proposals to reduce nutrient pollution largely by providing cash incentives for farmers have not been funded by the Bush administration and Congress. A new so-called '*action plan*' is scheduled to be signed next week in New Orleans," but Turner said he doubted it would have much impact 'there's no promise of funding,' Turner said. "Ron Litterer, president of the St. Louis-based national corn growers Association, said the last two years have seen a higher yield of corn but that the amount of runoff was reduced. He said farmers and seed companies alike are using techniques to produce higher yields using less acres, tillage, and nutrients."

Other solutions hinted at

At a later point in this book, I will go into an extensive plan to trap and utilize the nutrients closer to the site of application. The idea will fully qualify in the area of carbon sequestration. In discussing the potential healing factors for the anemia of the Gulf Stream, I have not properly stressed two almost wholly natural factors. They are ash fall from volcanoes and from forest fires. Timing wise and for nutrient content they do not wholly replicate the once regular (seasonal dust raising, ocean micronutrient supplying) migrations of Buffalo's. The other factor I submit is not quite natural—but perhaps the red soils of Arkansas, Oklahoma, and Texas have already made their positive contributions. Let's consider first the history of iron and phytoplankton growth: from HTTP: //EN.Wikipedia. org/wiki/ocean fertilization

Iron and Phytoplankton—but don't forget Sprengel's Law.

"Consideration of iron's importance to phytoplankton growth and photosynthesis dates back to the 1930s when English biologist Joseph Hart speculated that the ocean's great 'desolate zones' (areas apparently rich in nutrients, but lacking plankton activity or other sea life) might simply be iron deficient. Little further scientific discussion was recorded until the 1980s when oceanographer John Martin renewed controversy on the topic with his Marine water nutrient analyses. These studies indicated it was indeed a scarcity of iron micronutrient that was limiting phytoplankton growth and overall productivity in these desolate regions, which came to be called 'high nutrient, low chlorophyll' (HNLC) zones.

Martin's famous 1991 quip at Woods Hole oceanographic institution, 'give me a half tanker of iron and I will give you another Ice Age' drove a decade of research whose findings suggested that iron deficiency was not merely impacting ocean eco-systems, it also offered a key to mitigating climate change as well. Martin hypothesized that increasing phytoplankton photosynthesis could slow or even reverse global warming by sequestering enormous volumes of CO2 in the sea. He died shortly thereafter during preparations for Ironex 1, a proof of concept research voyage, which was successfully carried out near the Galapagos Islands in 1993 by his colleagues at Moss Landing Marine laboratories. Since then nine other international ocean trials have confirmed the iron fertilization effect."

Because this rendition of Sprengel's law was commended to Congress early in George Dubya Bush's eight-year tenure as president and *apparently* ignored by the Congress, (possibly for reasons of political expediency, they probably couldn't determine the prime beneficiaries nor did they have the means to tax them). I'd like to add my small bit of logic to the incentives for the Congress.

Like the soot fallout from the Kuwait oil fires and the pumice from Mount Pinatubo, the differential surface heating of selected areas of the ocean will have their pronounced onshore and downwind Effect. That rainfall should it reach California's central valley in time to enhance springtime orchards and truck garden's, (or more reasonably perhaps, ease the present deficiency of winter snowpack in the Rockies) could enhance the quality of life (by virtue of re-raining) for all of water short western United States. It might even compensate for the structural deficiencies built into car crazy Los Angeles, who led the world in the mania of road construction, which contributed to the artificial dryness so beloved of Southern California moviemakers. There are many other places in the offshore world which would line up to beg for the specific ocean largess if it were planned and attracted correctly. Perhaps the cold runoff from Greenland could be properly fertilized for someone's onshore benefit but the Atlantic situation is very much in flux so I would defer to future studies derived from the future Lake Qattara. But the water deficit now and in future for the Mediterranean cries out for this kind of phyto-plankton enhancement.

The need for and limitations of iron enhancement

Now back to the scholars quoted previously: "about 70% of the world's surface is covered in oceans, and the upper part of these (where light can penetrate) is inhabited by algae. In some oceans, the growth and or reproduction of these algae is limited by the amount of iron in the seawater. Iron is a vital micronutrient for phytoplankton growth and photosynthesis that has historically been delivered to the pelagic sea by wind driven dust storms from arid lands. This aeolian dust contains 3 to 5% iron and it's deposition has fallen nearly 25% in recent decades due to modern changes in land use and agricultural practices as well as increased greening of dry regions thanks to increasing levels of atmospheric CO_2. (Arid zone grasses and vegetation now lose less water vapor through their stomata to absorb the same amount of carbon dioxide, and thus stay green longer, reducing dust storm frequency and the amount of iron reaching the deep seas. Increasing sand desertification does little to compensate for the shortfall since sand is primarily silica with relatively low iron content . . .)

"In 'desolate' HNLC zones, therefore, small amounts of iron (measured by mass parts per trillion) delivered either by the wind or a planned restoration program can trigger large responsive phytoplankton blooms. Recent Marine trials suggest that 1 kg of fine iron particles may generate well over 100,000 kg of plankton biomass. The size of the iron particles is critical, however, and particles of 0.5 to 1 micrometer or less seem to be ideal both in terms of sink rate and bio availability. Particles this small are not only easier for cyanobacteria and other phytoplankton to incorporate, the turning of surface waters keeps them in the euphotic or sunlit biologically active depths without sinking for long periods of time."

CARBON SEQUESTRATION—deep ocean "plankton that generate calcium or silica carbonate skeletons, such as diatoms, Coccolithophores. And foraminifera, account for most direct carbon sequestration. When these organisms die their carbonate skeletons sink relatively quickly and form a major component of the carbon rich deep-sea precipitation known as marine snow. Marine snow also includes fish fecal pellets and other organic detritus, and can be seen steadily falling thousands of meters below active plankton blooms.

"Of the carbon rich biomass generated by natural plankton blooms and fertilization events, half or more is generally consumed by grazing organisms (zooplankton, krill, small fish, etc.) but 20 to 30% sinks below 200 m into the colder water strata below the thermocline. Much of this fixed carbon continues falling into the abyss as marine snow, but a substantial percentage is redissolved and re-mineralized. At this depth, however, this carbon is now suspended in deep currents and effectively isolated from the atmosphere for centuries or more. (The surface to benthic depths cycling time for the entire ocean system is approximately 4000 years.)" Authors note. I don't wish to appear impertinent but the abyss of knowledge shortfall rears its ugly head once again. *This reminds me of the kind of thinking that propelled radioactive wastes in cement filled drums to the 'benthic depths' in whose tranquil solitude they were supposed to remain for thousands of years—but emerged in a few years—corroded and defiled and defiling!!!*

Dimethyl sulfide—cooling—But where?

One more paragraph may prove to be helpful with my additional commentary." Dimethyl sulfide and clouds. Some species of plankton produce dimethyl sulfide (DMS), a portion of which enters the atmosphere where it is oxidized by hydroxyl radicals (OH), atomic chlorine (CL) and bromine monoxide (BrO) to form sulfate particles and ultimately clouds. This may increase the albedo of the planet and so cause cooling."

Probably it is too much to hope for to have a proximal Deepwater location upwind from the Sahara Desert where specific forms of dimethyl sulfide producing phytoplankton could enhance evaporation and cloudiness where prevailing winds could spread it over the Sahara as a double-barreled asset for the arts of the rainmaker but the timing sequences are known and most of the potential liabilities are also **KNOWN**.

By the by, the Antarctic Ocean is renowned for its nutrient wealth and it also has an apparent excess of atomic chlorine above it. I don't necessarily recommend Antarctica as a rainmaking site, certainly not a first choice. But when we come to monitor and regulate the unwieldy buildup of ice or of liquid water . . . . Remember now that the conservative choice for Antarctica implies no sudden changes. The remedial choice would seek to alleviate that recent buildup of excess heat around Antarctica. We have already noted the natural reluctance of evaporated water to fall on excessively heated sand.

Here I am offering only one proof for this admittedly nebulous area of causation, and that is the spectacularly sudden increase in South Atlantic ocean heating from nothing exceeding 74° to large regions of 80+ culminating in the first-ever South Atlantic hurricane striking Brazil in 2004.

All of this foregoing material is but a prelude to justify the exorbitant cost of the exit canal from Lake Qattara. The timing of this exit water has to suit the affected northern hemisphere nations and their choices merely highlight the global consequences of all of our choices.

But the main purpose is to bring moisture and hope to the nations around the Mediterranean

We'll explore the consequences for neighboring nations. (There goes the neighborhood! That's a joke!!! Remember, "Reverence for life!") Under the heading of water sources and there are and should be multiple sources. For primary guidance I refer you to the chart of freshwater stress for 2002 and projected for 2025. From the Encyclopedia Britannica Yearbook 2004 page 192." In 2002 the stressed areas include Libya and Egypt but stretched eastward to include Saudi Arabia, Iraq, Iran, Afghanistan and Pakistan," in short the troubled areas of the world. The anticipated stress zone for 2025 expands to include all of North Africa and eastward to include India and north and east of the Caspian Sea

SPECIAL REPORT

World Water Crisis
Is There a Way Out?

by Peter Rogers

"Of all the social and natural crises we humans face, the water crisis is the one that lies at the heart of our survival and that of our planet Earth."

Such was the dismal state of the world's water supply, as presented in a press release by Koichiro Matsuura, director general of UNESCO, on March 5, 2003. Matsuura later warned, "Over the next 20 years, the average supply of water worldwide per person is expected to drop by a third." For years there had been warnings of an ever-worsening crisis in the availability of water on planet Earth, and in making 2003 the International Year of Freshwater, the UN gave the issue global prominence. The signs are troubling. Rapid rates of population growth worldwide, rapidly growing income in many countries, and consequent rapid urbanization have led to highly stressed water systems. (See MAP.) It has been estimated that 2.3 billion people live in areas where there is not enough water available to meet basic needs of drinking, sanitation, hygiene, and food production—defined as 1,700 cu m (2,200 cu yd) per person per year. Some 1.7 billion people live under true water scarcity, where the supply is less than 1,000 cu m (1,300 cu yd) of water per person per year. Under conditions of scarcity, lack of water begins to hamper economic development as well as human health and well-being. All of these troubling signs are magnified by the possibility that we may be entering a period of rapid human-induced climatic change, with very uncertain implications for water-resource management in the future.

In 2000 the UN General Assembly set a goal "to halve the proportion of people without access to safe drinking water by the year 2015," and in 2002 the UN World Summit on Sustainable Development approved a supplementary goal of halving "the proportion of people without access to basic sanitation." The UN estimates that 1.1 billion people do not have access to safe drinking water (defined as meeting minimal standards of bacterial and chemical quality) and that 2.4 billion people do not have adequate sanitation. Cutting these numbers by 50%—while at the same time increasing food production, reducing poverty, and sustaining the ecosystem—is an ambitious goal. Hasty, ill-conceived responses may only exacerbate the problem, and so the best response at this time may be to think

FRESHWATER STRESS

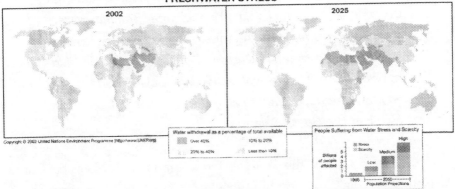

In a bit of whimsy, it's plain to me that the chart maker included India knowing that any love affair (anticipated) between India and the motorcar would produce soil hardening and Ocean filming. But it is also clear that any addition of freshwater either evaporated and re-rained or any of five other methods of supply should be welcomed throughout the neediest water short zone on the civilized planet. Allow me one more passing reference, to the incompleteness of the Gibraltar plan. How much of the increase of Mediterranean salinity is due to the late-season flooding and downpours that are mentioned repeatedly in the "late cyclones of data from the Mediterranean?" It is certainly not mentioned! By way of homespun comparison, a 6 inch downpour in 1951 in Nebraska produced a half-mile wide railroad upsetting cataract out of a river Big Blue that was little more than a wide creek ordinarily. As much as one third might have been non-H2O in that muddy expense, but who thinks to ask what percent of that awesome sight is salt?

The proper perception of teleconnections should draw down the Atlantic at the prime time to revive the Agulhas Current and draw that 80 degree water from around Antarctica up to the Equator to revive the Sahel—Sahara!

One more footnote to a nearly dead plan: Maybe the excess concern about the contaminated Mediterranean could be dissolved by reviving the moribund Agulhas Current in its efforts to bring cleaner Pacific water to the contaminated Atlantic. My plan for revived evaporation within the Qattara Depression should help to bring about this revival, as well as moving toward the equator the newly enhanced and mostly unwelcome hot water in the South Atlantic.

It's time to lay out the sources for water in a more orchestrated fashion. First of all, the primary tunnel from the Mediteranean supplies primarily surface sea water to the depression. Even the opening of the power generation complex should be attended by a delicate sense of timing. The first goal of water extraction from the med should not interfere with the already scarce natural rainfall pattern. If enough sophistication is achieved, the second goal should slow the explosive and erosive nature of the late-season cyclones by initiating the rainfall cycle earlier in the season. To help to achieve that level of sophistication, the tunnel should be built with a high intake and a low exit near the water intake at the beginning of the tunnel. Two different grades of salinity will have ready access and egress to and from the depression. The third water resource for the Qatar depression should be seasonal. Tanker loads of Antarctic water carried in such a timely way that their cooling effect contributes to a sophisticated weather system around Qattara. These tankers will hold their rinsing until they have carried the freshwater to the Qatar depression. That rusty oil will then no longer contribute to the hesitation of ocean evaporation, and will be fully remediated in a climate zone where the result of cool water, increased rainfall, and raw material for the 'coming of the green', will fully constitute a triple blessing.

The third goal should awaken baffled incomprehension in 93% of my readers. It will become more evident as we discuss additional sources of water for the depression and the cumulative effect of low albedo green and darker wetter soils. That goal is to enhance a more regular and cleansing, nourishing pattern of rainfall throughout the med and beyond the borders of Lake Qattara in the broader Sahara. This includes the 'natural' recharge of the depleted 7,700 sq mile wetlands around the Niger River in West Africa. To be introduced Shortly!

Dual intake and timed hot exit from Qattara

The two-stage intake for the Qatar depression is not adequate for that goal. It will require broader assistance than even the regulated exit for heavy, hot saline water. To justify the complexity of this next **threefold addition of water to the Sahara I must** remind you that not only was the Sahara once green but the enhanced irregularity of water vapor caused in part by an occasionally oily ocean is sufficient to recommend the compensatory expansion of evaporation and remediation zones in the Qattara, the Dead Sea, Death Valley, and perhaps the Salton Sea but the special circumstances prevailing there will require the highest of expert consultation before any plan can succeed. Afar Depression and the Qatar depression are ideally equipped to receive the oily residue of the 5000 tankers that ply the world's oceans. *Death Valley could with monumental effort be prepared to receive the floating excrescences (read, tarry seepage that some sources suggest equals Exxon Valdez every four years) that contaminate the Los Angeles basin and probably inhibit the already scarce natural rainfall. They already have the tar eating flies from the La Brea Tar Pits.*

The Dead Sea already has quantities of asphalt that crop up on occasion. But another area nearby is perhaps ideally equipped to receive the oily residue of tankers and that is the Afar Depression. I've supplied data on Afar earlier but didn't establish its corrective connection to the dried up and diverted swamp called Bahr el Jebel. **In other words, Afar can rebuild the earlier trigger of swiftly rising moist air columns that can hopefully arrive at the North Pole to trigger earlier Atlantic evapo-transpiration.**

I'm going to name a couple of extra assets for water to the Qatar and Afar, but it will be several pages before the best brains in NASA can begin to translate them into an effective plan. First and foremost, water from Antarctica to the Qatar depression. The primary supply should come from a single hulled VLCC ship carrying the precious icy liquid from beneath Antarctic ice that now threatens to inundate the coastal areas of the planet before the frightened population can adapt or move and very secondarily possibly by Russian icebreaker and tugboat, in short an iceberg. The logic of the icebreaker may become apparent when the water under Antarctica is mined. The logic of the tugboat should become apparent as the climate inputs of cold ice are recognized and valued in a timely way, *i.e. mated to the timely arts of the rainmakers.*

The Niger should be the prime beneficiary of South Atlantic, 80 degree waters.

That's two but we haven't begun to account for the increase of 80° heat accumulating around Antarctica and in the South Atlantic. That heat must and will have an outlet. If it only expends itself as hurricanes on Brazil's eastern coast or on additional ice deposition on the Antarctic continent, adding top heaviness to Antarctica's frozen assets, it will of course enrich an already top-heavy media with an abundance of juicy scare headlines.

If we augment the seasonal evaporation from the med at the same time as the reawakening of the anemic Gulfstream in its ability to carry heat and moisture to the far north, it then becomes feasible to hope that excess Antarctic heat will dissipate northward drawn by a reduced level of Atlantic sea level. If the process needs a singular event to be reinitiated, a massive early or late season hurricane on US shores (mostly retained) could be of long-term catalytic benefit: but you don't see me or anyone else recommending it.

> *I'd much rather do it by accelerating, in a timely seasonal fashion, the long term evaporability of Mediterranean water and thus enhance its freshness by drawing from either the Agulhas or the Amazon spring flood.*

Needless to say the long-term heat gain around the Antarctic could already be affecting springtime flood margins for the Amazon, the Congo, and even the Nile. Of the three the Nile should be most prepared to utilize the renewal of Antarctic heated water.

Introduction to the Niger River

But there is a fourth river that stands in the path ready to receive the largesse of a newly activated South Atlantic Ocean. A brief introduction is in order. From the Internet HTTP://US dotf838 mail.yahoo.com/ym /s howletter?Msgld=3564_3309206_4926_2493_47088_0_61_12416 1_1928192038ldx+0 . . . Pg. 16 Of 31 German advisers help African nations sort out water needs (the Niger River).

"The Niger is an intractable River. First, it flows hundreds of miles northeastward into the interior of the African continent. Then, all of a sudden, it makes a bend and starts to flow in a south easterly direction. Finally it flows into the Atlantic on the coast of Nigeria. Altogether the river crosses for countries on its 2600 mile journey: Guinea, Mali, Niger, and Nigeria.

The water from Africa's third longest river is in great demand in all four countries, which is why negotiations regularly have to be held about the trans-boundary use of the river. Among others, Peter Pieck of the German agency for technical cooperation (GTZ) helps to ensure that these consultations run smoothly. He works at the office in Niamey, the capital of Niger.

'Currently, three large dams are planned along the river—in Guinea, in Niger, and in Mali, says Pieck. These plans raise a large number of questions. How large should the dams be? How much water will the dams let through? What should have priority—energy production or environmental protection? The supply of drinking water and the simple disposal of wastewater also have to be clarified.'"

First of all, while I'm screwing up my courage to the sticking point, let me say with generous approbation, these are wonderful projects.' Having said this much, there are two parts of this manuscript which may provide helpful guidance to the dam planners. First of all the Niger is in the direct path of that massive amount of newly stored heat around the Antarctic ice cap. When, (not if) that extra heat vent's, the dams had better be adequate to more than the average historically predictable stress. Second, the new science of re-raining should induce a proper prediction of the downwind effects of the newly stored and later evaporated (hence available for re-raining) water.

If these concerns are properly weighed and worked into the planning process, the foundation will be laid for the enduring prosperity for West Africa and a positive influence ranges eastward.

Back to the raw data: "the Niger is not only a special river because of its unusual course, but also because of its inland delta, which stretches across an area of more than 7700 mi.^{2} in Mali. This region has huge wetlands with enormous species diversity, says Pieck. 'Sometimes the River thins and almost dries out. In years of extreme drought it has even been known to totally disappear. The constantly fluctuating water level is a permanent worry for local farmers. However, the supply of electricity from the hydroelectric power stations in the dams is decisively affected by the amount of water the Niger carries. These two criteria—the irrigation landscape and electricity generation—are very frequently at the center of the negotiations of the ABN, which was founded in 1980. Whether the individual national governments decide water management issues or ABN's executive Secretariat is allowed to have a say is an important question in this context—and holds considerable conflict potential." But back to the Nile—the same paper is helpful.

"Another major project of the German agency concern is the Nile, Africa's longest river. The agreement on the distribution of the waters of the Nile is currently being renegotiated between all the neighboring countries. Consultations are long overdue, as the old treaty still stands from colonial times. The river's total available volume needs to be ascertained as the basis for these negotiations. however this data has never been collected." *(I don't thoroughly trust the Nile data that I cited earlier in this text. But I do trust the description of the man-caused changes in world climate that I covered for the last 85 [now 88] years.)* "Now, the GTC is supporting the development of appropriate databases that do not only cover the Nile and thus are laying the foundation for future water projects in Africa." [Author's interjection: If the renewed rainfall from the recently heated Antarctica water should become too aggressive for Egyptians in Cairo, or the new reservoirs planned for the upper Nile, I'd recommend diversion to the prehistoric route of the West Nile, which would help us add to the resources of the Qatar depression. But that diversion, is only an insurance policy against three possible future developments.

The first is the improbable influx of galactic cosmic rays and their extra nuclei and cloud forming capability. Obviously no consensus can form quickly enough to provide an effective insurance policy for Dec. 21, 2012.

The second concern is the long-term heat gain around Antarctica. Whether it comes from the 13,800 km² of newly deiced Ocean, extra ultraviolet penetration, or the short sharp closeness of the Antarctic

Sun, during the Antarctic spring due to the motion of the Malenkovitch cycle. It is a genuine concern and should be viewed as a resource.

Third, there is not enough continent in the southern hemisphere to dissipate it south of the equator. It probably cannot be trained to dissipate onto Australia; but it might be retrained to fall once again on to the Sahara desert. In order to facilitate the development of the proper size (colossal) consensus, I include the following material from the book, Floods, Droughts, and Climate Change, by Michael Collier pages 50 and following. "The Sahel desert (which includes parts of Senegal, Mauritania, Mali, Burkina Faso, Niger, Nigeria, Cameroon, and Chad) continues to be gripped by a drought that began 40 years ago. Malam Garbe, a farmer in Dalli, Niger, remembers villagers hunting antelope, monkeys, wolf, fox, squirrel, and rabbit, before the drought. Now there is no game. Forty years ago, he and his brother grew enough millet to feed their families and provide a surplus for sale; now they cultivate three times as much land but harvest only 1/7 as much grain. Sand dunes lap at their windows as dry wind strips away their soil and hope.

From north to south, the Sahara region of western Africa was anywhere from 37 to 15% wetter than the long-term mean throughout the good years in the 1950s. Crops flourished then. Beginning in the early 1960s and persisting through the present, rainfall abruptly decreased: 31 to 13% below the long-term mean in the 1970s and 24 to 20% below normal during the 1980s. The consequences have been dire. This region, verdant enough to export grain at midcentury, is now haunted by famine, disease, and the dislocation of untold thousands of inhabitants. To make matters worse, the Sahel's population has risen from 60 million to 207 million during this drought." 12(Danish,K.W.,1995, international environmental law and the 'bottom-up' approach: a review of the desertification convention, Indiana school of Law {HTTP://www. law.Indiana.EDU/GLSJ/volume 3 number one/Danish.HTML}.

Sahel now compared to Dust Bowl of the 30's

"What mechanisms could produce this change? First and most obvious is the clearing of 123,000,000 hectares of natural vegetation for agriculture. Native perennial vegetation was replaced by crops such as millet, peanuts, and sorghum. Once a farmer depletes his field of nutrients, he simply abandons it and plows up new land. As trees are removed, the wind moves at ground level **like a scythe, grasping at the vulnerable topsoil." (Sounds to me like an identical description applied to the 'dustbowl' of central USA in the 30s).** "The director of Niger's national department of the environment estimates that an area the size of Luxembourg is lost each year in his country to this degradation process called Desertification.

> *"The bare bones of the Sahel offer a stark lesson in the anatomy of a decades long drought. Much of Africa has seen periodic ups and downs of rainfall throughout the last 40 or 50 years. But the Sahel appears to be caught by some sort of feedback mechanism that prevents its recovery from the short-term droughts that only briefly affect the rest of the continent. Desertification could presumably change the color and tone of the Earth's surface and therefore change its reflectivity or albedo. Reduced albedo should lead to reduced soil moisture, convection, and rainfall. Realistic modeling of albedo and soil moisture changes, however, fails to fully account for long-term drought in the Sahel."*

An African Civilian Conservation Corps

"A great deal of interest is now focused on dust introduced into Africa's lower atmosphere by wind. The reddish dust rises in stupendous quantities on winds called harmattan's that have darkened the sky for ages." **(Does it not bear a great resemblance to the "black rollers" of the American Dust Bowl that were slowed down by massive tree planting by the Civilian Conservation Corps and of course the cooperative efforts of hundreds of thousands of farmers!?)**

WE NEED A CIVILIAN CONSERVATION CORPS FOR THE UNITED STATES OF NORTHERN AFRICA. Hopefully this becomes the catchphrase signaling the beginning of a brand-new USnA.

They could understand their mission as a supplement to the Kenyan Nobel prize winner, the lady who claims 30 million trees for last year and wants to go for 1 billion in 2008. But without the magic of timely water and careful planting of hardy trees (and groundcover) it will be for naught. Now back to Michael Collier.

Chapter12

SPECULATIVE CAUSE AND POTENTIAL FOR CURE OF EL NINO

"'Blood rains' from the Sahara fell on Portugal and Spain in 1901; an estimated 2,000,000 tonnes of red mud coated Southern Europe in April 1926. Escalating rates of soil disturbance can only inflate these figures. Dust rising to 5000 m significantly accelerates local atmospheric heat absorption. The high-level African easterly jet stream normally carries rain bearing clouds to the Sahel from July through September when the ITCZ (intertropical convergence zone) swings to its northern apogee." *My speculation follows in Boldface underlined italics: And does that unscalable mountain of hot dusty air reach its acme [forming a dam] at the same time as the easterly trade winds lose their ambition to pile up water in the Western Pacific, thus becoming one of the so far unnoticed triggers to the return of Pacific waters, which in a few short months is labeled El Nino with its speculative fascination for climatological specialists from all portions of the globe? I will offer some more extensive speculation on the causes and reduction of El Nino, but it probably remains technically outside the ethical scope of this argument because right at this moment we still believe that we [humans] didn't start it*

"The location of both the ITCZ and the African easterly jet stream depends on regional north-south temperature gradients. When dust is present in sufficient quantities, temperature gradients can shift sufficiently to displace both the jet stream and the IT CZ's moisture bearing storms away from the Sahel. If this relationship withstands scientific scrutiny, it would be a link connecting at least one type of decades scale climate change to the actions of humans. Are we comfortable thinking of ourselves as an 'external force' that can so fundamentally change the world in which we live?"

I must confess at this point that I am uncomfortable even with my strictly limited ethical goals to simply right the wrongs that have been committed over time by men, civilizations, and/or corporate entities. *But the similarities of the Great Plains Dust Bowl with the Sahel, propel me forward in spite of the massive size of the Sahara and the length of its deprivation. Two additional difficulties present themselves. One—the Coriolis effect works a bit better to drive Atlantic moisture West. Two—the rain shadow of the Himalayas!!*

But we've made a proper beginning with two saline choices of water from the Med to the Qattara Basin. Are you still holding your breath for the breathtaking water possibilities? Additionally we have the option of three possibilities for Antarctic cold (water and/or ice) to be selectively introduced to the Qattara. Ice, cold water and early season ultra-abundance of flood waters for the Congo, the Nile and the Niger should be factored into the ultimate goal of attracting water and wind to work their cooling magic upon the newly heated South Atlantic. But all of this water sourcing is gobbledy-gook unless we establish the central purpose of these potentially conflicting water sources. *They are in short, Catalysts for the renewal of the real-thing-tropical rainfall that actually reaches the ground—triggering the lowered albedo so necessary for the regreening of that 3,000,000+ square miles of mostly overheated sand.* The third concern is born of success. Yes, we are that 'external force',(both oily oceans and excess civilized hardening coupled with bad farming on the negative side) but this proposed change can pose all of the problems of a brightening future. We have warm oceans and that's an asset! We can seed with iron filings but the seeding will be followed by localized (surface as opposed to deep) warming, and the location and time need to be honed to be in the path of the shoreward bound breeze. We can cool the sand of the Sahara even beyond the confines of Qatarra. We can locate moist seedable air currents above the Sahara. But we still need to form the team which can plant the groundcover to stabilize the sand long enough for the trees to take root. The choice of groundcover tree types, (remember the succession of tree types which produced the towering sequoias?) The teams need to be closely coordinated with the rainmakers and where successful need to be closely followed by more seeding. That includes both choice and timing of plant seeds which must be followed closely by cloud seeding to stimulate plant growth. One additional line of guidance for the planters—the sparse natural rainfall for the Sahara apparently comes toward the end of summer.

Neither that line nor the sophisticated flying of Hans Ahlness qualify for the best guidance for a Chapter which I will call **Rainfall for the Sahara Desert. Hans may have been trying too hard to produce rain that had already been milked out of the skies by the Rocky Mountains and/or overheatedly suppressed by the hot dam of air hovering intimidatingly over Southern California. If you love conflict and ultimately resolution this is your section of the book. Be advised that you may soon be gasping for air in the watery deluge. But many sober-minded people are well advised to go on to other portions of the book which will appear to be within range of commonsense reality.**

RAINFALL FOR THE SAHARA DESERT

Before I can complete your introduction to the modern science of rain making, I feel it necessary to indicate some of the complexity of the assignment for the rainmakers. Here is a brief look at the Quasi-Biennal zonal wind Oscillation (QBO).

Andrew Heaps, William Lahoz and Alan O'Neill Centre for Global Atomospheric Modeling Department of Meteorology University of Reading UK. The discovery of the QBO "The eruption of the **KRAKATAU** Volcano" (6° South 105° East) On August 27th 1883 lead people to believe that the stratospheric winds above the equator blew in a westward direction. Dust from the eruption took 13 days to circle the equator and this upper air wind became known as the **KRAKATAU** easterlies.

In 1908 Berson launched observational balloons above lake Victorian in Africa and found westerly winds at about 15kl (120mb). These westerly winds are called Berson's Westerlies.

These conflicting results were resolved through the work of Reed (1961) and Veryard and Ebon (1961), who showed that the wind above the equator oscillates in direction. It was shown that the wind in the stratosphere changed direction on average every 26 months and that the alternating easterly and westerly wind regimes descend with time.'

We are only examining this interesting data for any helpfulness to the rainmakers and I have to confess there is not much help. No data on the humidity is available and the probability of significant moisture being available in a descending air stream is next to nil.

But the visible evidence of the 10 ft. icicles chopped from the balloon basket remains as a concrete form of encouragement.

*Let's move on to another form of modern rain making.
This one is a bit closer to the moist breezes of the
Pacific Ocean.*

The standard model of rain making is best revealed in an article titled "Reining In The Weather" By Donovan Webster. "Not Far From The Dead Dog Saloon, Behind a Body Shop On the Main Street of Grantsville UT, Stands a rusting, 4ft tall metal box. The box sits atop a box of gaseous iodide. That, when fired up sends a plume downwind towards the nearby Oquirrh Mountains. Once carried up on the wind each silver iodide crystal forms a core, or nucleus, around which water droplets collect. Since silver iodide has a crystalline structure similar to that of ice, it allows the tiny water droplets to coalesce until they are big and heavy enough to fall out of the sky, ultimately increasing snow fall between 10 and 15% a year. That's more water for later release across the state's thirsty desert during spring and baking summer. More water for irrigation, livestock, human consumption, and sports. It means millions of dollars in water related revenues for the state's economy every year.

The Utah cloud seeding effort comes courtesy of North American weather consultants, America's oldest weather modification company, located in an upscale office park in nearby Sandy, Utah.

Founded in the 1950's" (Apparently Charles Mallory Hatfield didn't incorporate . . .), "The group is currently run by two solid citizen scientists with commercial aims, Don Griffith, and Mark Solak, who have spent their careers working in privately funded weather modification efforts around the country and the world.

"In Colorado they seeded the Gunnison River drainage, a series of reservoirs and dams in the west of the state. In California they run seeding programs for the Santa Barbara water agency, a group that says the effort may increase rain in target areas up to 20% a year.

In reality cloud seeding is pretty low tech. A tank of silver iodide topped by a burner and surrounded by a perforated-metal wind arrester.

The whole contraption is hooked to a tank of propane to provide the flame and warmth that lifts the silver iodide into the atmosphere.

'We've got lots of cloud seeding units in mountainous areas all around Utah,' Solak says.

When wind, temperature, and humidity are just right the company calls local residents who are paid a fee to go out and turn on a cloud seeding unit, sending a plume of silver iodide downwind why an array of cloud seeders? Although a single plume cannot change the world, a group of such seeders, each responsible for a small shift in precipitation, can often tilt the balance locally, driving rainfall or decreasing the intensity of storms.

'In weather modification, the uninitiated think you must make huge impacts on the atmosphere to get a desired result.' Griffith says. But it's actually the opposite. If we just make tiny modifications to existing conditions, little touches here and there, the changes then cascade upward using existing weather's natural actions, and that's what gets the biggest results.'"

Further hints that NASA should be in charge of the overall timetable.

Before we turn this job (planning and coordinating) over to an agency of proven competence for oversight purposes—(the masterful introduction to this group of people will come later, courtesy of Don Trauger's book "HORSEPOWER TO NUCLEAR POWER." Lest the job be considered too elementary for the carefully honed minds and spirits of this group of people, I'm going to embark on a lengthy discourse which is in part frankly speculative and perhaps dangerously outside of my ethical license. Which is: Unlimited 'messing' with the climate is outside of my ethical scope,

WE'RE ONLY ENTITLED TO FIX WHAT WE HAVE AGREED IS BROKEN, AND AT LEAST PARTLY OUR FAULT.

Since this material can only complicate the reasoning process of the 'fixit-experts', I HAVE ALREADY DARKENED THAT PORTION OF THE MANUSCRIPT WHICH MAY BE UNIQUELY MY OWN SPECULATION. Nevertheless we need to have a solid foundation of fact before any speculation can take root. For that I turned again to the book." floods, droughts, and climate change", by Michael Collier and Robert H. Webb. Pages 53 and following . . ." air and water differed greatly in how they acquire and transmit heat. Unstirred water conducts heat 27 times faster than air. At sea level, water has a heat capacity 3500 times greater than an equal volume of air. The oceans, therefore are vastly more energetic than an equal volume of air. The topmost 10 m of an ocean typically holds as much energy as the entire atmosphere above it." (This is a starter version of a complex scenario which may suggest regulation, to some is to suggest that hurricanes are merely a visible symptom of the distress **of the ocean which is forced to store more than its normal amount of heat.**

Further attempts to distinguish between a gentle spring rain and a constipated late season towering hurricane.

(This rendition doesn't say much about the heat dissipation rate of the extensive undersea volcanoes which are only now beginning to be catalogued.) {Whether that heat is stored deep in the ocean by sunlight [including UV], coming through cloudless skies, released by undersea volcanoes, or its release is retarded by layers of impenetrable oil [spills, waste, motor oil, or natural seepage, is not quite immaterial. It might even lend some material to the mostly vanity argument whether "global warming" is man caused or nature caused*! Whatever the cause, the excess of storage leads to the extremes of weather so beloved of headline writers!!)*

The cure for hurricanes-at least the big bad late season Atlantic ones-is contained in this simple paragraph. Lush, luxuriant tropical vegetation producing early season mountains of rising air that cools quickly enough on its way to the poles to dislodge the abundance of stored cool air occupying less than normal space because of plentiful wintertime precipitation. That cool dry air should it happen to march down the Atlantic Corridor, will harvest and drive ashore the moisture of the Atlantic which if left undisturbed till later in the season will inevitably rear itself to the colossal height and fury of the Hurricane.

I'm well aware that this rendition almost completely ignores the reality of El Niño, and La Niña, which have fascinated climatologists for years. Let's consider that from the same source. For the collected facts and ideas I'm indebted to Michael Collier. Anything lost by abridgment is my culpability. My ideas which piggyback on top of his research are a bit like a flea hitching a ride on a grizzly bear and perhaps imagining that me (the flea) is in charge. (Re; Floods, Droughts and Climate Change pg 52)

"Energy is not distributed homogenously within the oceans, neither vertically nor horizontally. The surface waters at most latitudes are warm relative to deeper waters for a simple reason: surface water directly absorbs incoming solar radiation; once warmed, this water expands, becomes slightly less dense, and tends to 'float' on denser, Deeper water. This process is so effective at thermally segregating seawater that most oceans have a clearly demarcated plane, called the thermo cline, separating their surface and deep waters. At the thermo cline, which might lie anywhere from 100 to 1000 m beneath the surface, water temperatures suddenly plummet by 10°C or more over a further descent of a few tens of meters. Water above the thermo cline can warm and cool seasonally, but Deepwater will have a temperature, usually 4°C or less that is remarkably stable over periods of hundreds of years."

Don't Ignore undersea volcanoes

So far as I can ascertain, the comforting stability of 4°C or less, does not fully grasp the heat dissipation pattern of the massive number of undersea volcanoes that are only now beginning to be catalogued. Additionally, this material does not capture and place the depth or temperature that the saline submergence of the Gulf Stream and the exit material from Gibraltar might choose to travel. The length of scholarly observation of undersea volcanoes may not be great enough to provide a rational base to hazard a good guess as to 'real' versus 'potential' volcanic increase driven by the 2012 'Galactic plane crossing'. And we shouldn't forget it's threatening corollary, the potential for the magnetic flip of North to South. Elsewhere in the manuscript, I have more or less casually referred to the need to sequester a billion tons of soot. The now obvious need for the soot is to forestall the precipitous return of snowball earth. It should be obvious to even a bureaucrat that it should be stored near the equator, and an airstrip, in flyable packages, for northward distribution.

A belated but telling defense of Al Gore.

This is probably as good a time to defend Al Gore as any. This winter's abundant snow has made him appear vulnerable and in some eyes misguided and dead wrong. But the excessive snow is merely the logical consequence of an ocean that is forced to harbor more heat than is normal. As long as we have "oily oceans" we will continue to run the risk of a more than average cold winter. If you accept my version of used oil being the culprit then the correct policy choice has already been made in most of America—to recycle used oil. But that policy is not complete while rusty tankers are plying the oceans and silently cleansing their holds in the handiest available manner. Even if we have the four major depressions of the equator made available for cleansing and remediating the oceans we may still have need for ocean dumping of oil during a time of excessive volcanism to partially inhibit the wintertime onslaught of precipitation. By now you may conclude with me that the soot bombs are a better way to encourage the earth to accept normal solar gain—This should be a decisive NO to snowball *earth! The further question that should be asked even tho' there is no obvious answer is this! What kind of soot? And will the micro-nutrients (or toxins) be friendly to ocean phytoplankton.*

IF Mr Gore Places an acceptably large bounty on the floating plastic in the Eastern and Western Pacific Garbage Dumps and neutralizes the toxic effect in the food chain and treats the resulting stabilized plastics as soil stabilizers for the shifting rootless sand of the Sahara, He will have my vote for the man of the century.

Proper help for the Soot Bombs.

In John Petersen's charmingly brief book, "A Vision for 2012" p.93, pub. 2008, under the heading of Policy Priorities, "Within each departmental area there should be a number of priorities that drive everything else. They should include: The environment: . . . find an effective method for carbon sequestration; . . ." on the next page, "Fund development of carbon scrubbers for coal fired power plants and mandate immediate installation on all such facilities . . ." Even though I may have been overexposed to CO2, as a teamster, I'm still not jumping on the—Bury Carbon at all Costs-bandwagon but as a future safeguard to the viability of our insurance companies we actually need Carbon Banking. We still need more facts. (Back to Michael Collier.) "The surface heating also varies horizontally, with temperatures generally increasing as one approaches the equator. Surface temperature becomes disproportionately important as water warms beyond 25°C. Above this temperature, H2O is more inclined to exist as a vapor rather than a liquid. A warm band of tropical water encircles the earth, lying within 20 to 25° latitude of the equator. Because of the Earth's tilted axis, the band shifts to one side of the equator or the other, depending on whether summer happens to be in the northern or southern hemisphere. Evaporation from the tropical oceans liberates a vast quantity of water that is carried into the atmosphere. The tropics are always studded with thunderstorms and regularly breed hurricanes and cyclones. Hot moisture laden air rising from the tropics enters the atmosphere and spreads north or south of the equator. Cooler air must slide back in along the surface—converging from North and South—to replace that which has risen within the ITCZ.$_{=1}$ Philander, S.G.H . . . 1996 'Why the ITCZ is mostly north of the equator.' **Journal of Climate. V. p. 2958-2972** . . .

Submarine Lava flows. Men probably didn't Cause them but they can upset the climate practically single-handed. "once this oceanic soup begins to boil, it is quickly stirred . . ." *Enough of these fundamentals for now. Since I'm going to dwell in praise of his work for some time—it seems appropriate to indulge in a little criticism. Prior to his stellar chapter 8 "the Christ Child" which I will excerpt extensively, he has a chapter titled, "oceans and air." His charts of ocean currents are probably as good as any that I've read. But only a single line, P. 39 "scientists have explored the possibility that large submarine lava flows (at least 10 km³ in size) may have occasionally heated ocean water enough to upset the world's climate patterns."*

The 1976 Guinness book of world records, page 122" the total number of known active volcanoes in the world is 455 with an estimated 80 more that are submarine. The greatest concentration is in Indonesia, where 77 of its hundred and 67 volcanoes have erupted within historic times." That's the old view. Robert W. Felix in 2005, pages 133 and following of his book" Not By Fire But By Ice," writes" Marine geophysicists aboard the research vessel Melville recently discovered 1133 previously unmapped underwater volcanoes about 600 miles northwest of Easter Island." (Easter Island is about 2300 miles west of Chile in the South Pacific.)

"And they're huge. Some of the newly found volcanoes rise almost a mile and a half above the seafloor. Even then their peaks remain about a mile and a half below the water surface. Consisting of both seamounts and volcanic cones, they're packed into an area of 55,000 mi.², about the size of New York State . . . 'It's the greatest concentration of geologically active volcanoes on Earth.' (Seattle Times, 14 February 1993) . . . only 5% of the ocean floor, said Mike Donald, has ever been

mapped in detail . . . . Plumes of water heated to almost 800° spout into the inky blackness."

A footnote goes on." These words were written in 1994. Today, scientists at NASA estimate that there may be as many as 1 million submarine volcanoes. As many as 75,000 of them rise a half mile above the seafloor and several thousand of those, in turn may be active. See HTTP://volcano.und. Nodak.edu/vwdocs/"

Rising oceans and hyperactive volcanoes which comes first?

It seems to me both presumptuous and premature to map ocean currents without reference to any currents which may be somehow related to the gigantic flow of undersea volcanic heat . . . biblical . . . we know in part, we prophesy in part . . .—but he still gets an A from me for effort. Let me use these facts to support a leading question. 'If you include all volcanoes within 200 miles of shore with all the newly catalogued and not yet catalogued volcanoes in the ocean, what percent of all volcanoes are oceanic or ocean influenced. My tentative answer is a guess—somewhere in the high 90th percentile. Only Yellowstone eruptions every 600,000 or 700,000 years and our own Missouri bootheel earthquake of 1811 come to my mind to appear to defy the oceanic tidal-weight-flux connection. My point is a bit simpler —GRAVITIC FLUX (moon and sun) is an important factor. I only say that to provide a logical governor on any water or ice extraction which may be done from the Antarctic ice cap for the benefit of lake Qatarra and its environs. The volumetric measurements of Greenland and Antarctic ice are almost accurate enough (separate from warming hysteria) to guide future extraction, even in the likely event that Greenland ice resumes its upward momentum.

A beginning look at El Nino which passes irregularly over the top of those recently counted 1,133 underseas volcanoes and others not yet enumerated.

Let's continue with the theme of presumption—this time mine— we are working toward El Niño the height of presumptive speculation—again mine. But with the able help of Michael Collier. p. 39 "El=Chi chon erupted in 1982, just before the greatest El Niño of the 20th century."

(And now it's time to interrupt with a potentially retractable compliment related to being grateful for the elephant in the kitchen. Should we be grateful for the predictive prescience of the Ixtoc I oil spill, of 1979? And did that spill shut down ocean evaporation enough so that the world-wide cooling of El Chi chon did not start us toward a full-fledged Ice Age? I doubt that any politician could be prescient enough to achieve that goal, but it's almost certain that he would have claimed credit for averting a worldwide catastrophe if he had.)

"Was the subsequent worldwide atmospheric temperature drop causally related or merely coincidental? Without the El Chi chon interruption, would the 1982-83 El Niño have been an even warmer event? The summer of 1816 was remarkably cold, but it was within bounds of known climate variability; again, was this caused by a volcanic eruption, or was it mere coincidence? Scientists at the University of Washington calculated that large eruptions such as Tambora decrease worldwide air and sea surface temperatures for a year or two. They could, however, find no [substantial] climatic evidence that such volcanic explosions have a demonstrable effect on worldwide precipitation patterns or sea level pressures that are integral components of El Niño events. Because climate change is so fundamentally tied to energy transport, it is tempting to try to relate it to volcanic activity. For the most part, however, this has remained an elusive connection."

My grandma and my grandpa's tales of the blizzard of 88 remain firmly stuck in my mind—it's not an elusive connection. Ocean sealing from 1978 on and albedo changes both from surface oil and from ash fall provide an explanation of transparent clarity. The deeper heating under an impervious oil blanket and the contrasting extreme surface heating of an ash-clogged or soot filled top layer bodes ill for the ocean's ability to regularly supply the continents with rainfall. The apparent paradox claimed by University of Washington scientist's captures the initial cooling of the ocean surface caused by atmospheric murkiness but, may disregard?. The immediate solar gain of the murky top foot, or so, of the soot laden, or ash clogged ocean, later producing deeper ocean coolness. The 1993 Mississippi Missouri flood was preceded by 2 1/2 years of soot fallout from the Kuwait oil fires (that was never factored in to my limited knowledge). Even though the Pinatubo eruption of 1990 was estimated for its element by element contribution to the atmosphere, my limited reading did not reveal a writer who drew a two or three year time line to the resultant ocean murkiness and subsequent Missouri Mississippi flood of 1993. The question needs to be asked. Was this a man-caused El Niño? But, it will not be answered right here. The limited speculation that there might have been four El Niño's between 1992 and 1998 is bizarre enough by itself. Chalk this up to my provincial sense, but that's my candidate for the greatest El Niño. But in my albedo colored view of the Earth's climate, the connected data of those years would constitute nearly conclusive evidence of either the greatest or most destructive experiment by mankind. The quality of the designation is reserved for the future use of the knowledge gained. In print it has been both ignored and labeled as an unheard-of five-year El Niño—some portions of that wonderful and alarming ocean heating probably didn't qualify for the title, "El Nino," since its major strength was not expended on Peru.

If we simply ignore the broad ramifications of this material: we can draw the narrowest of possible conclusions. Oil or soot, or volcanoes or vegetative murkiness or increased ultraviolet, can combine in bizarre ways to enhance the erratic nature of our oceans capability to supply the continents. This residual knowledge was tapped into recently to argue for the safety of the oil drilling platforms because their safety record in oceanic storminess was an effective insurance policy against the (harmful) effects of natural seepage. Therefore the answer given *in the 1947 Reader's Digest," big damn foolishness" remains valid, but with the carefully punctuated addition of timely transpirational releases with the goal of reducing the extended dryness on the shoreward side of the climate extremes. This includes the 500 gallons cisterns proposed for the corners of the urban—suburban housing in settled areas where excessive rainwater flushing has increased the water deficit.*

El Nino at some length.

Since I am examining El Niño, not for its global effects, but for the effects that the anticipated changes to the Sahara and Sahel might produce on this major wildcard in the weather—climate scenario. A gentle transition to Michael Collier and El Niño is in order. From page 48: in addition to Milankovitch cycles and sunspots—external forces." There are internal forces as well, which are best exemplified by the much ballyhooed phenomenon called El Niño." Several excellent examples follow . . . from pages 64 and following "what is El Niño? That's easy: El Niño is a coupled oceanic and atmospheric phenomenon that occurs every few years(roughly every 3 to 7), with ramifications that ripple out from the equatorial Pacific to touch the lives of people from Peru to Canada to South Africa. El Niño is the quasi-periodic fluctuation of equatorial Pacific water temperatures and a coupled atmospheric response: warm water sloshing back and forth across the basin, with winds rising and falling in unison. You might think of this innate oscillation as the climate system unconsciously drumming its fingers. The devil, of course is in the details. Perhaps it's best if we begin with the snapshot of the 'normal' Pacific Ocean stretching from Peru and Ecuador, west to Indonesia.

The general flow of water in the southern Pacific Ocean is dominated by the South Pacific Gyre, a counterclockwise drift driven by global winds and shaped by the Coriolis effect. The eastern limb of the Gyre includes the Humboldt and Peru currents, which flow north past Chile and Peru. These currents are sustained by persistent winds blowing northward along the western coast of South America. As wind pushes the water, the phenomenon known as Ekman transport, caused by the Earth's spin, dictates that flow will be redirected to the left of the wind track in the southern hemisphere. Thus surface water moves away from the coast. Deep cold waters rise to fill the void. The thermocline—that sharp delineation between warm surface layer and cold deep water—is usually very shallow in these upwelling regions, typically 40 m or less. Surface mixing processes reach deep enough to bring up cold water from beneath the thermocline. The take-home message: Pacific ocean surface water along the coast of South America is normally cold.

Peruvian fishermen, accustomed to these nutrient rich cold waters that normally upwell along their coastline, had long ago learned to recognize a current of warm water that was likely to change fishing prospects about Christmas time each year. Some years, perhaps one out of every 3 to 7, the current would be dramatically warmer and the fishing would be much poorer . . . cold upwelling waters along the Peruvian coast, disrupted by the annual incursion of a moderately warm current from the north, is the 'normal' picture in the eastern equatorial Pacific. Now let's turn our attention to the western Pacific Ocean near Indonesia. This region is directly influenced by persistent easterly trade winds that converge at the equator from both the north and the south.[2] these winds drive equatorial currents that pile Western Pacific water into a mound 60 cm high.[3] the winds tend to diminish as they reach Indonesia; the Southern Current slows down there and the water heats up, frequently exceeding 28°C. The thermocline is a couple hundred meters below the surface.[4] the take-home message: the Western Pacific is typically a bulging pool of very warm water.

We need one additional element to complete the simplest of models: atmospheric circulation. Waters of the western Pacific approach the limits to which an ocean can absorb solar energy. Above 30°C, evaporation occurs at such a furious pace, and heat is transferred so quickly into the atmosphere, that the ocean cannot long sustain higher temperatures. This air expands and rises, carrying prodigious amounts of both heat and moisture into the troposphere. Much of the rising air moves north and south from the equator in the Hadley cells described earlier. *But a significant fraction also flows laterally away from the Western Pacific either farther west or back east in a zonal (as opposed to a meridional, or North South) flow. Air flowing eastward at upper tropospheric altitudes descends upon the eastern Pacific. Trade winds on the surface complete the loop back (a fruitless loop) to the west.*

"Cold water dominates the eastern equatorial Pacific, and warm water dominates the West, and dry air descends in the East, creating an area of high pressure. This system is held in dynamic equilibrium between solar heating and the Earth's attempts to redistribute that heat. The equilibrium drifts back and forth across the equator annually as the ITCZ tracks north or south and the sun moves from winter to summer . . ." a brief history of the term El Niño is in order that you may comprehend the nuances of the change that I will propose. From page 67 and following *"Gilbert Walker . . . asked to direct the Indian meteorological service in 1903 . . . [was] charged with predicting the tropical cyclones that ravaged British ships in the Bay of Bengal. Soon the service was also expected to predict droughts that decimated India when it's precious monsoons failed to materialize. More than 4.5 million people had died in the drought and terrible famine of 1899 . . . .*

Gilbert Walker knew little about meteorology when he first showed up but he was scientist enough to carefully gather data from specific locations throughout India and simultaneously take a very large view of worldwide weather patterns . . . Walker looked at snow cover in the Himalayas, rainfall over the Nile, and atmospheric pressure records from Australia to Chile.".(If there had been a system of atmospheric pressure recording for the entire Sahara desert, it is my guess that he would have included it profitably in his theories.) *"What he most wanted was to create a reliable system for predicting the Indian monsoon.*

Along the way, he stumbled upon a persistent pattern of oscillating atmospheric pressure; first low pressure in northern Australia, and then low-pressure farther to the east. The oscillation seems to seesaw back and forth every few years. The Southern oscillation is now standardized as the pressure difference between Tahiti and Darwin, Australia. Walker knew something, but his correlations of the Southern oscillation with the Indian monsoon remains less than reliable . . ."

Jacob Bjerknes, University of California at LA . . . laid the detailed groundwork for understanding zonal air movement over the equatorial Pacific . . . Walker circulation (named) in honor of Sir Gilbert . . . (brought) evidence linking the Walker circulation with rhythmic changes in sea surface temperatures across the Pacific. He noted that, as easterly trade winds decrease, the Western Pacific's warm bulge begins to relax and flow eastward toward South America. With warming of the Pacific Ocean near the international date line, more evaporation occurs there and more energy is pumped into the atmosphere above western South America. The trade winds collapse or even locally reverse, further accelerating and sustaining this breakdown of the Walker circulation. Monsoons in India and Australia give way to drought. The annual El Niño current that flows southward along the South American coast becomes noticeably stronger and warmer as Ecuador and Peru are hammered by unusually heavy rain storms and floods.

This coupled ocean atmosphere process has become known as the El Niño—Southern oscillation (ENSO) phenomenon . . . it's multi-year rhythms suggest that forces beyond volatile atmospheric dynamics must dictate its waxing and waning., certainly oceans with their huge capacity to hold heat, have enough thermal momentum to direct this multiyear periodicity, but why is the process not more predictably cyclic? . . ."{The author of this manuscript is attempting to answer that question in three ways, on the ground, in the air, and in the sea, but I'm not going to condense the answers into three simple catchphrases just yet.} Back to Michael Collier, who is still leading my thoughts. But please don't blame him for the conclusions that I leap to!

"The answer to this question remains elusive. One approach involves backing into the problem by asking a related question: has the Pacific Ocean always experienced ENSO events? . . . Jeffrey Dean's tree rings record ENSO's that occurred as much as 2600 years ago. Ice-core and coral records show that the ENSO machinery has been steadily chugging along for at least the last 2000 years.

"The scientists have extended their vision back 15,000 years in at least one location by examining debris flows that swept into a lake high in the Andes of Ecuador. Radiocarbon dates were obtained from organic material within laminae of mud and gravel. Going back in time, the scientists could document wet dry cycles with a periodicity of 2 to 8.5 years from the present—until about 5000 years ago. Even farther back a change becomes evident. Before 7000 years ago, the debris flows were less frequent, occurring about every 15 years . . . The implication is that a clear ENSO-like signal began to develop 7000 years ago and was not fully recognizable until 5000 years ago."

The roots of El Nino are unmasked

"The only way to find the limits of the possible is to go beyond them to the impossible." Arthur C. Clarke. It is not nearly enough just to note that the timespan for the beginning of the ENSO-like signal coincides with the birth of the Saharan sandpile. I would like to note also that the easterly trade winds are the 'normal'. To counter that 'normal', requires more than a mild-and-temporary deficit in the low-level returns of cooler air to the equator. *I believe it requires a 'double damn' of hot air 'mountains', to constipate the normal flow of cooler air from the poles back toward the equator, to fulfill their primary mission—carrying sea generated moisture onto the continents surrounding the ocean. The first 'mountain', is described by Michael Collier in the "harmattans" that rise from the floor of the Sahara. The second was described as*

an 'invisible mountain' of hot dry air that centered on Amarillo, Texas in the summer of 1980, leading to Kansas City's hottest summer in recent memory.

Fanciful thoughts about re-watering the Sahara—except it worked in the Great Plains Dust Bowl to produce the incredible surpluses of the late forties and fifties.

The missing data needed to bring this argument to a suitable level of sophistication would be extensive studies of the variation of timing of the color change (albedo from white to vegetated absorptiveness.) Of the Himalayas and the drama of the timing of the change of color of the Eurasian landmass from snow white to absorptive wet darkening. This could be done as a 'compare and contrast to other seasons with lesser albedo contrast, (as in sandy and grayish green with its accompanying lesser UV absorption—because drier.)

Nevertheless, I am convinced that one could appropriately hope for a reduction in the severity and frequency of El Niño events if sizable parts of the Sahara could be returned to their previous level of grassland and trees. You remember from a few pages back that Jennifer laid out the probable sequence for the advent of the giant Sequoia's . . ." first Arctic grasses and shrubs were the only plants that could take hold in the barren soil . . . when they died they decomposed, and over thousands of years a layer of topsoil built up . . ." {Authors note} 'I don't think that American impatience can wait even 100 years for soil buildup. Less than 15 years was the turnaround time for the decimated soil of the Great Plains. Even shorter was the rehab time for the newly engineered right-of-way of the interstate system thanks to the soil stabilizing plastics spun out of the petroleum barrel, that and common bales of wheat straw!'. The United States Department of Transportation has a catalog of soil types, grades, types of vegetation and success rate of each application. I suspect that a great deal of that knowledge could transfer successfully to the rehabbing of the Sahara desert: but I'm not sure that it would be thoroughgoing enough to transfer the water holding capacities of that 'stationary Camel of Arizona', the Saguaro cactus, to portions of the Sahara desert.

CAN A FULLY VEGETATED SAHARA (above) ENHANCE AND REGULARIZE THE TRADE WINDS THUS MINIMIZING THE GIANT TRAGEDY OF EL NINO CITED BELOW?

Lest I be accused (likely by me) of lightheartedly playing Don Quixote and tilting at Saharan Camels, allow me to add some 'gravitas' from other sources. From "the National Geographic magazine, El Niño/La Niña natures vicious cycle," by Kurt Suplee. Mar 1999 pg. 73ff.

It rose out of the tropical Pacific in late 1997, bearing more energy than a million Hiroshima bombs. By the time it had run its course eight months later, the giant El Niño of 1997-98 had deranged weather patterns around the world, killed an estimated 2100 people and caused at least 33 billion {U.S} dollars in property damage . . . Christmastime—El Niño . . . that titanic storm source would pour vast amounts of precipitation onto Peru's normally arid northwestern coast. But few have ever seen this much rain—5 to 6 inches a day in some places.

"The runoff from the floods poured into the coastal Sechura desert. Where there had been nothing but arid hardscrabble waste for 15 years, suddenly—amazingly—lay the second largest lake in Peru: 90 miles {145 kilometers} long, 20 miles {30 km} wide, and 10 feet {3 m} . . .

"Changed weather . . . Indonesia and surrounding regions suffered months of drought. Forest fires burned furiously in Sumatra, Borneo, and Malaysia . . . Temperatures reached 108°F {42C} in Mongolia; Kenya's rainfall was 40 inches {100 cm} above normal; Central Europe suffered record flooding that killed 55 in Poland and 60 in the Czech republic; and Madagascar was battered with monsoons and cyclones. In the US mudslides and flash floods flatten communities from California to Mississippi, storms pounded the Gulf Coast and tornadoes ripped Florida . . .

By the time the debris settled and the collective misery was tallied, the devastation had in some respects exceeded even that of the El Niño of 1982 83 which killed 2000 worldwide and caused about $13 billion in damage.

And that's not the end of it. It is not uncommon for an El Niño, winter to be followed by a La Niño one—where climate patterns and worldwide effects are, for the most part, **the opposite of those produced by El Niño. Where there was flooding there is drought, where winter weather was abnormally mild, it turns abnormally harsh. La Niña's have followed El Niño's three times in the past 15 years—after the 1982 83 event and after those of 1986-87 and 1995. Signs of another La Niña began to show up by June 1998."**

More 'gravitas'to follow: but there are limits to how much bad news the human psyche can tolerate. I am going to attempt to put some positive spin on the above and then supply a chart from the National Geographic (September 08 page 131) that I choose to interpret in a hopeful manner.

A GREEN MOMENT

Pollen studies show grasslands and pockets of forest periodically flourished in the Sahara. Many factors, including a wobble in the Earth's axis, helped shift seasonal rains northward.

8,000 years ago

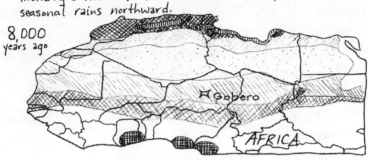

TODAY

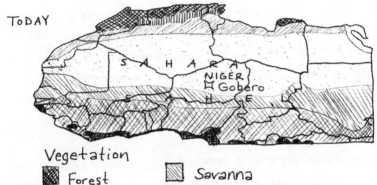

Vegetation

- [Forest] Forest
- [Steppe] Steppe
- [Semidesert] Semidesert
- [Desert] Desert
- [Savanna] Savanna
- [Savanna and forest] Savanna and forest
- [Rain Forest] Rain Forest
- [No Data] No Data

drawn from National Geographic, September 2008

This next sentence is too lighthearted even for me! You'll note 8 lines back—' where winter weather was abnormally mild, it turns abnormally harsh —. Just bring on La Niña and presto-winter is back. My interpretation lacks scholarly credentials but perhaps only a farmer who has prayed fervently over a field of crops—just two days from being seared irrevocably by the burning Southwind—(Santa Ana—Black roller—harmattan, et. al.—can appropriately appreciate the drama of the change from desert drab to growing (absorptive) green. Even the silvery yellow wheat stubble can turn absorptive green with the miraculous addition of timely rainfall in hours or days (but of course we've been trained to call them weeds and only recently has the 'no-till' philosophy allowed them to remain. We've already noted the astonishing superabundance of El Niño moisture amid normally desert like environments. The next observation can only be in the form of a question!! Does it really need to take 3 to 7 years for this equatorially lofted 'green air' to die off in its North Pole arrival time and subsequent re-turn as the backbone of the trade winds hence allowing the reconstitution of yet another El Niño. That, in a nut shell, is the 'in the air' portion of El Niño causation. We'll get to the pivotal land causation factor, shortly. This air factor only explains one side of Michael Collier's beginning account of El Niño. 60 cm of "hot water slipping back east from Indonesia to Peru.'

A "wanna-be-farmer's fervent thoughts require us to probe further into the present reality of El Nino studies

Now let's go back to the Sahara desert, and gravitas! This time from the same National Geographic page 130." Scientists long suspected that the Sahara wasn't always dry. Then during a 1981 shuttle mission NASA used a synthetic aperture radar to peer through the sands to the desert's underbelly, revealing a network of river courses inscribed in the bedrock during multiple Wet intervals." That map, detailed earlier, creatively reveals a Sahara of 8000 years ago and the Sahara of today. It's not much of a stretch to claim that the Sahara and the Sahel of today as a desert area almost tripled the size of yesteryear. In my terms that equals "hot damn" in spades!!

Now, 'gravitas with a bit more authority and even with a trace of optimism. From HTTP: //www.pmel.noaa.gov/tao/faq.html 8/13/2008'

Selections:" frequently asked questions about El Niño and La Niña . . . Theory or fact? El Niño as a physical occurrence is a proven fact . . . . We do not have(thunderstorm prediction quality data) knowledge for El Niño.(onset!)

"Once an El Niño has started, we have reasonably good skill in predicting the subsequent evolution over the next 6 to 9 months, but before it has started we have very little skill in predicting the onset before the event becomes obvious. There are a variety of theories for why El Niños start, but none of them has given us real skill in making a forecast in advance, the way we can for thunderstorms."(The science advances faster than I can record it! I'm betting that the next El Niño will be predicted more than six months in advance.)

It already happened! One source claims to go back to the 1800s with the limited data then available to successfully predict (via 20-20 hindsight), the advent of El Niño. While we are on the subject of undue ocean heating it is a short step to El Niño. But unlike the seeming randomness of oceanic volcanism El Niño is potentially predictable at least in my eyes and curable. Maybe this is too bold a choice of words, perhaps subject to amelioration. Back to the facts.

" It must be said that there is still plenty of social utility in predicting the evolution of an El Niño after it starts, since that gives six months or so warning before the effects come to the US. For instance, a fairly weak El Niño started earlier this year, and that enables forecasters to predict that the coming winter is likely to be warmer than normal across the northern states, and wetter than normal along the Gulf Coast. Such forecasts are certainly useful for farmers and water managers, but from a scientific point of view they are unsatisfying because they do not answer the fundamental question of why the event started in the first place.

"One reason for this state of affairs is that El Niño's only come along every 4 to 5 years or so, so there aren't very many to study, (we've had decent instrumentation in the tropical Pacific for less than 20 years)."

I have to insert what must sound like carping! They have successfully counted 1133 underseas volcanoes in that area; but there is no active monitoring of them for duration and totality of heat output worse yet! With all of the eye in the sky satellites surveying all portions of the globe, nobody (to my knowledge) has been assigned to total the unbroken series of sunshiny days that are almost certainly the necessary prelude to El Niño. And if you count them do you reckon them as surface heating due to flotsam, soot, fallout, or nature's phytoplankton: or is it going to be deep heating (including ultraviolet) of an ultra-transparent Ocean? That probably calls for more instrumentation and there are at least $33 billion reasons to improve the monitoring (and potential alleviation of damages). back to normal gravitas.

"Thunderstorms happen every day in summer, so there's been lots of opportunities. In the case of El Niño, one theory is that these events are the means by which heat is drained from the equatorial oceans after a period of accumulation. Such a theory predicts that by observing the growth of heat content, it should be possible to forecast when an El Niño will occur. That seems to be at least partly true, but it was (apparently) contradicted by the El Niño of 1993, which occurred immediately after one the previous year, and no (apparent) accumulation had occurred . . .

"El Niño's are marked by simultaneous westerly (from the West) winds, warm SST (sea surface temperature) and deep interface, and occurred in 1986, 1991 92, 1993,(the weak one I mentioned earlier), 1994-95, 1997-98(an extremely strong one) and you can see the present one developing. HTTP://www.pmel.noaa.gov/tao/jsdisplay/Click "assorted plots" button, then pull down the menu and click "Monthly EQ UWND SST 20C ANOMS"

The peak of presumptive speculation 5 El Nino's but you must know that I was nearly stranded in the middle of the rampaging Missouri in 1993, so cut me a little interpretive slack!

Remember that in most cases the manuscript bold face underlining points to personal speculative interpretation: here goes. In an eight-year period four El Niño's and no theory? Here's my theory! Kuwait oil fires. 130,000,000 gallons at least. Soot fallout monitored at Hawaii stations—no adequate record of depth and or duration, or extent.

Second, Pinatubo, Philippines—time and volume elsewhere in the script. No record of the duration of the floating residue. The only reasonable clue we have as to its shallow warming, from surface murkiness, and its corollary subsequent deeper cooling— is the distinct pause after 1998—to (8-13-08?), before the next El Niño.

The mysterious presence of the 'inexplicable' 1993 El Niño subsides quickly when some of that hundred and 30,000,000 gallons of flammable Kuwait oil is factored in as a fallout. But my point is that the human factors which contributed to bunching five El Niño's in an improbably short eight years could point to a very human strategy for minimizing (not eradicating) El Niño. But the absence of a following La Niña might become the proper focus for further studies since we are properly concerned with cooling (selected) parts of the globe. Here is more on the impossibility of eradicating El Niño from the same source of 'Gravitas'.

"Information contained in the chemical composition of ancient tropical Pacific coral skeletons tells us that ENSO has been happening for at least 125,000 years. This span of time covers the last Ice Age cycle when the Earth's climate was cooler and very different from today's climate. In addition we can reasonably assume that the ENSO cycle has been operating ever since geologic processes closed the Isthmus of Panama about 5 million years ago to form the modern boundaries of the Pacific basin."

It seems a bit of a stretch to take the hot and cold evidence of coral skeletons from a hundred and 25,000 years to 5 million years. But the closing of the Isthmus of Panama suggests to me a link to the (supposed by me at least) increased evaporation capability of the Atlantic Ocean. We remain properly focused on enhancing its regularity. A second note as to the coral skeletons suggests to me a monumental difference between the flooding and dryness of coral skeletons at or near sea level and the extensive marks of flooding high in the Peruvian Andes.

Back to 'Gravitas' "there is nothing we can do to stop El Niño and La Niña events from occurring. The year-to-year oscillations between normal, warm, and cold conditions in the tropical Pacific associated with the ENSO cycle involves massive redistributions of upper ocean heat. For instance, the accumulation of excess heat in the eastern Pacific during a strong El Niño like that which occurred in 1997-98 is approximately equivalent to the output of 1 million medium-sized 1000 MW power plants operating continuously for a year. The magnitude of these natural variations clearly indicates that society cannot hope to consciously control or modify the ENSO cycle. Rather, we must learn to better predict it, and adapt to its consequences."

Anticipating Southern California's water choices!

Since we have already modified El Niño, (five in an eight-year period in the middle 90s sounds a lot like semi-deliberate human modification.) I suggest that we work on the quality of modification rather than stick our heads in the sand. One interesting idea that didn't come from me, appeared in the media with the profit motive firmly behind it: "the oil platforms in the path of hurricanes didn't leak a drop when tested by severe winds and waters— and the ocean already has numerous natural oil seeps— so why not drill?? We'll discuss one rather large natural Seep, when we consider Southern California's contribution to water adequacy for Southwest United States.

Let's return to quoted material:" are all El Niño's the same? Every El Niño is somewhat different in magnitude and duration. Magnitude can be determined in different ways, such as variations in the Southern oscillation Index SOI. Plots of sea surface temperature anomalies (SSTA) from the ENSO monitor, show El Niño's back to 1982, including the 1982-1983 El Niño, which until 1997, was the largest El Niño of this century . . . ."

Since we already have a clear perception of magnitude —$13 billion in 82-83, graduating to $33 billion 97-98. Not all of that increase is inflation, some estimates for 97-98 run as high as $100 billion. This upward tilt of El Niño is unacceptably bleak!

"Do El Niño events occur only in the Pacific Ocean? The great width of the Pacific Ocean is the main reason we see El Niño Southern oscillation (ENSO) events in that ocean as compared to the Atlantic and Indian oceans. Most current theories of ENSO involve planetary scale EQUATORIAL waves. The time it takes these waves to cross the Pacific is one of the factors that sets the timescale and amplitude of ENSO c**limate anomalies**. The narrower width of the Atlantic and Indian oceans means the waves can cross those basins in less time, so that ocean (Pacific) adjusts to wind variations more slowly. This slower adjustment time allows the ocean atmosphere system to drift further from equilibrium than in the narrower Atlantic or Indian Ocean, with the result that Interannual climate anomalies (e.g. unusually warm or cold Sea surface temperatures) are larger in the Pacific.

The landmass as magnet for oceanic evaporation!

"There is another way in which the width of the Pacific allows ENSO to develop there as compared to the other basins. In the narrower Atlantic and Indian oceans, bordering land masses influence seasonal climate more significantly than in the broader Pacific. The Indian Ocean in particular is governed by monsoon variations, under the strong influence of the Asian land mass. Seasonally changing heat sources and sinks over the land are associated with the annual migration of the sun. Heating of the land in the summer and cooling of the land in the winter sets up land-sea temperature contrasts that affect the atmospheric circulation over the neighboring ocean. This land influence competes with ocean and atmosphere interactions which are essential for generating ENSO."

My contribution. This is solid basic science information! The enhanced ocean—continent interaction as a preventative of El Niño needs further examination from this point of view. Under what conditions does the land mass act as a barrier (instead of an enhancement) to ocean—continent interaction? The question is answered pictorially elsewhere! And repeated below!

Gravitas again question 16 page 8 of the preceding material." What is the relationship between hurricanes and El Niño?

In general, warm ENSO episodes are characterized by an increased number of tropical storms and hurricanes in the eastern Pacific and a decrease in the Gulf of Mexico and the Caribbean Sea . . . Atlantic Ocean. It is believed that El Niño conditions suppress the development of tropical storms and hurricanes in the Atlantic; and that La Niña, (cold conditions in the equatorial Pacific) favor hurricane formation. The world expert in this area of study is professor Bill Gray of Colorado State University . . . Pacific Ocean . . . El Niño tends to increase the numbers of tropical storms in the Pacific Ocean."

I'd like to take a crack at spinning these truths, my way. El Niño in the Pacific augments the heat reception in the Atlantic (clear, cloudless, sunniness and heat storage courtesy of increased ultraviolet penetrating power makes the Atlantic the breeding ground for the giant storms of the future. And searching for the hidden speculative truth behind the mountain of data: it is wholly natural and human to focus on the storms. The more helpful focus is on the time between the storms—extended storage time of sunshine pouring down day after day uninterrupted by clouds or atmospheric murkiness. When that becomes the focus, the sowing of the ocean with iron dust and the specific breed of phytoplankton which contributes to upper atmosphere murkiness, (mentioned elsewhere) becomes a tool in the hands of the climatic manipulator . . . I don't recommend this because it is obviously outside of my ethical license—unless it contributes to the renewal of trees and grass in the heart of the Sahara desert.

Chapter 13

HADLEY CELLS THEN AND NOW

There is an almost invisible transfer of emphasis in the climatological writings that I see as I'm reading. I don't know whether it's real in anyone's mind besides mine. But I'll discuss it in the order that I encountered it. Hadley cells, supposedly a step-by-step—down transfer of equatorial energy toward the poles, rising and falling and eventually arriving at its lowest point near the North Pole. You'll remember that the original Hadley had the air traveling almost nonstop from the equator to the pole. The newest rendition describes Meridional transfer of heated (pre-El Niño) air going east or west from the Western Pacific near El Niño's starting point near Indonesia. My very important question: Is this the nearest thing we have to proof of North Pole warming in the past 35 years I.E., the resistance of the warmer poles to transfer of downhill air from the equator? Or does it measure the absence of the upward bound greened air from the recently denuded Sahel?

Same source, pg9 question 18. And "what is the relationship between greenhouse warming, El Niño, La Niña, and climate prediction?"

There is a lot of confusion in the public about the interrelations connecting climate phenomena such as El Niño, La Niña, and greenhouse effect is it true that a warmer atmosphere is likely to produce stronger or more frequent El Niño's? We don't know the answer to this question. It is certainly a plausible hypothesis that global warming may affect El Niño, since both phenomena involve large changes in the Earth's heat balance. However, computer climate models, one of the primary research tools for studies of global warming, are hampered by inadequate representation of many key physical processes, (such as the effects of clouds on climate and the role of the ocean). Also, no computer model yet can reliably simulate both El Nino and greenhouse gas warming together. So, depending on which model you choose to believe, you can get different answers. For example, some scientists have speculated that a warmer atmosphere is likely to produce stronger or more frequent El Niño's, based on trends observed over the past 25 years. However some computer models indicate El Niño's may actually be weaker in a warmer climate. This is a very complicated, (but very important!) issue that will require further research to arrive at a convincing answer."

A confusing paragraph in which authority—(Gravitas)-effectually ignores all of the elements I tried to discuss previously including the perviousness of the ozone layer, presence or absence of clouds, presence or absence of particulate matter and at what levels in the atmosphere (implying potential longevity), then the depth of the solar transmissions into the ocean. I especially would like to eliminate the phrase "global warming" as being too general (and conflict laden) and replace it with specific ocean oil sealing. It is far less than fair to use the term 'ocean oil sealing' when we may be referring to the chemical Pandora's box, derived during and after WW2 from the lump of coal and the petroleum barrel. We'll get a bit more specific in the following two pages when we discuss the Eastern Pacific Garbage Dump as well as the Western Pacific Garbage Dump. Much of this "carbon magic" is still floating and being partially consumed by those species of ocean fauna not yet facing extinction. When we transfer the harvesting skills from the great Plains combines to The Eastern and Western Pacific Garbage Dumps the resulting carbon reserve can be converted to a tarmac-like substance which when spread on Saharan Sand enables roots to stabilize and hold moisture in the useable top inches to expand the foundation for the resurrection of the Sahara. We know that many modern nets and ships have been idled in the Atlantic because of too many ships and nets and not enough fish! This last sentence has specially powerful bearing on the level within the ocean at which solar energy is converted to heat. This will happen because the life forms which can prosper and grow in the former dump will be energy storing rather than sterile and dead. The only missing element in this intriguing carbon recovery and helpful reuse is the size of the bounty needed to initiate the recovery process. These seemingly lightweight opinions can be transformed into heavy-duty facts. Let's go to Wikipedia: http://en.wikipedia.org/wiki/Great-Pacific-Garbage-Patch Like other areas of concentrated marine debris in the world's oceans, the Great Pacific Garbage Patch formed gradually as a result of marine pollution gathered by oceanic currents. The garbage patch occupies a large and relatively stationary region of the North Pacific Ocean bound by the North Pacific Gyre (a remote area commonly referred to as the horse latitudes). The gyre's rotational pattern draws in waste material from across the North Pacific Ocean, including coastal waters off North America and Japan. As material is captured in the currents, wind-driven surface currents gradually move floating debris toward the center, trapping it in the region. The patch's size is unknown, as large items readily visible from a boat deck are uncommon. Most debris consists of small plastic particles suspended at or just below the surface, making it impossible to detect by aircraft or satellite. Estimates on size range from 700,000 Square kilometer (270,000 sq. mi) to more than 15,000,000 square kilometers (5,800,000 sq. mi.){the larger of those two = 8.1% of the size of the Pacific Ocean} Or up to 'twice the size of the continental United States.' The area may contain over 100, million tons of debris . . .

An estimated 80% of the garbage comes from land-based sources and 20% from ships . . ."main article: Photodegradation The Great Pacific Garbage Patch has one of the highest levels known of plastic particulate suspended in the upper water column. As a result, it is one of several oceanic regions where researchers have studied the effects and impact of plastic photodegradation in the neustonic layer of water. Unlike debris which biodegrades, the photodegraded plastic disintegrates into ever smaller pieces while remaining a polymer. This process continues down to the molecular level . . ."Some of these long-lasting plastics end up in the stomachs of marine birds and animals, baby chicks, including sea turtles, and the Black-footed Albatross. Besides the particles'; danger to wildlife, the floating debris can absorb organic pollutants from seawater, including PCBs, DDT, and PAHs. Aside from toxic effects, when ingested, some of these are mistaken by the endocrine system as estradiol, causing hormone disruption in the affected animal. These toxin-containing plastic pieces are also eaten by jellyfish, which are then eaten by larger fish. Many of these fish are then consumed by humans, resulting in their ingestion of toxic chemicals. Marine plastics also facilitate the spread of invasive species that attach to floating plastic in one region and drift long distances to colonize other ecosystems."

I believe we have shown enough here to demand a clean-up no matter the cost. But the chemical consequences of a second use as a seed mat and soil stabilizer for the Sahel-Sahara must depend on the intelligent selection of rehabilitating plant material. See the later section on cattails. **Setting out the facts of the Eastern and Western Pacific garbage dumps. If Mr. Gore can set an effective bounty on this mess which results in the full and complete harvest and can use the magic of chemistry to detoxify it for a stabilizing plant foundation for Saharan transformation; he fully deserves my nomination for Man of the Twenty-first century!**

We have shown that murkiness in general can only hasten the arrival of El Nino and perhaps prolong it! All of this can be modified by the presence (or absence) of cool (cold) dry winds which have *their ultimately effective genesis in vegetation inspired equatorial air currents!! Nevertheless it provides an effective compare and contrast! Now back to gravitas.*

"Both 1998 and 1997 had record-setting global mean temperatures and also El Niño. What influences what?" I'd like to put this question on hold!

The answer will be revealed (below) in a satellite picture of the Eastern Pacific Ocean and Western United States. One picture is worth 1000 words, but I'll still try to explain what you're looking at. The cloud over the eastern Pacific is the rough equivalent of 1 million lost solar horse power. Western United States appears to be almost totally cloudless. When you consider that the standard motion of air above the 30th parallel is from west to east, you then have visible proof of the 'Hot Dam' effect of overheated land surface. This paragraph and picture are an essential introduction to the consideration of water adequacy for Southwest United States.

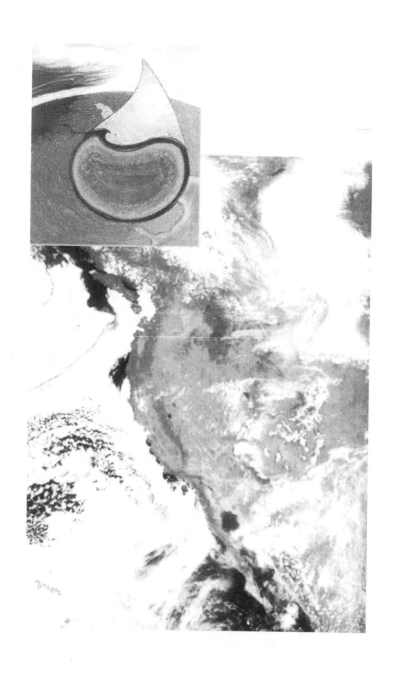

<< STRATOCUMULUS OFF CALIFORNIA

Sheets of stratocumulus clouds occur readily over the Earth's oceans. They play an important part in the Earth's energy budget because they reflect a large amount of solar energy back into space. This huge cloud bank is packed against the Californian coast, anchored there by the cool ocean beneath. It is unable to spread inland because much higher land temperatures break it up. Taken in March 2002, it is possible to see the affects of spring on the San Joaquin Valley (green area, top left), while snow still covers parts of the Sierra Nevadas.

Now let's transfer to a different source of gravitas: "the Encyclopedia of hurricanes, typhoons, and cyclones." 1998 edition page 288 (after a short discussion of many zones of above average SST {sea surface temperatures} the author documents a 10° lowering following the passage of a hurricane "while not every tropical cyclone that has either strengthened or weakened has done so because of an encounter with changing sea surface temperatures, there is sufficient meteorological data to establish that in passing over a particular stretch of ocean, several mature stage tropical cyclones have significantly lowered sea surface temperatures, in some cases by as much as 10°. Studies conducted in 1977 on hurricane Anita revealed that sea surface wind speeds within the category four systems eye wall had swiftly churned the first 100 feet of the Gulf of Mexico into deep radiating swells; 100 feet beneath that, the hurricane's mounting inverse barometer developed a series of upward spiraling currents that drew cooler water to the surface in its wake. Sea surface temperatures recorded following Anita's passage were 8° below what they had been 24 hours earlier . . . meteorological satellites equipped with infrared technology measures sea surface temperatures through the use of an instrument called a radiometer."

"These dramatic temperature reductions should effectively highlight the albedo changes which occur naturally within the time span of the hurricane's cloudy duration. Now we shift to another dramatic albedo change which our 'gravitas' author gives short shrift. Question 20. "Is it feasible to haul icebergs from Antarctica to the tropical Pacific to cool down El Niño?

The answer is 'NO.

"The simple reason is that to cool the tropical Pacific down to its normal state once an El Niño is underway would take an amount of ice 10 m thick covering an area equal in size to the continental United States. That's a lot of ice and there is no way to extract and transport that amount of ice with existing technology. Even if it were technically feasible, it would in all likelihood cost an astronomical amount of money, many times over the combined global losses due to El Niño.

"Furthermore, it would take a long time to transport. The inevitable delays that attend any grand project would probably mean you'd get all the ice to the tropical Pacific just as the El Niño was ending. It would be too late to do any good. Of course, since El Niño is often followed by La Niña, (which has its own set of adverse consequences on weather), you could end up exacerbating the effects of natural climate variability on society."

Let's think outside the box for a bit! One hurricane can drop the ocean temperature by 8 to 10° in as few as 1 to 10 days. We've already noted that clouds can provide a reflectivity of 90%. I've heard of a berg as large as Delaware (2200 mi.²). We have no notion of the duration of the berg, but the captain of the Titanic probably didn't have a reasonable estimate of the lifespan of the berg that destroyed his great ship. I suspect the author of what I've chosen to label as 'gravitas, is thinking of the cooling effect of melting ice rather than the much greater effect of heightened albedo. My own simplistic *reference to cloud—making returns to the lake effect snow of 11 feet derived from the passage of the Northwest wind over Lake Erie and the disastrous result for Buffalo at the east end.*

That situation poses one end of the cloud cover duration problem in its starkest simplicity. To research the other hand, (highflying, long-lasting clouds,) we might have to do time travel back to the late 30s, 40s, and 50s . . . some potent clouds were still with us in the 60s and 70s but we didn't notice their absence until the skin cancer rate began to multiply in the more pristine air of the 80s.

For memorable proof of the cooling effect of hurricanes, I would like to cite the October gale, hurricane of October 3, 1841. From the "Encyclopedia of hurricanes, typhoons, and cyclones," 1998, David Longshore, page 232." 85 men and boys . . . perished in the storm . . . (which dropped nearly 18 inches of snow over Central Connecticut." In a " globally warmed world", a hurricane that takes tropical moisture at 80+ degrees and converts it to 18+ inches of snow by the 3rd of October has to qualify as a 'good' hurricane versus 'bad' hurricanes. Neither the Saffir-Simpson scale, nor the dollar amount, nor the total of lives lost qualifies as a good hurricane with me! Only the term tropical storm rises to the occasion. That is why I spent time discussing the cloud making effects of cosmic rays and the selective outbreathing of certain types of phyto-plankton for a milder, gentler transition to rainy weather. No one (to my knowledge with the possible exception of Michael Collier {previously quoted} has linked the vigor of the Tradewinds to the less traumatic formation of weather clouds but I doubt that it would be a blind avenue for future research!! Back to gravitas.

"Finally the extraction of that much ice would seriously damage the environment of Antarctica. It could also have potentially serious consequences on global climate if it led to changes in surface reflection of sunlight, or had other effects on land surface processes . . ." *this, in my estimation is a careless finish to an otherwise outstanding presentation. Antarctica has been changed! One of the man-made changes, (12,800* km of newly ice free ocean) has been projected to continue for most of this new century,{courtesy of the enhanced ultraviolet energy coming through the man-made ozone hole}. Documented disappearances of the food chain and certain species of Penguin effectively highlight the crisis; but the critical feature of South Atlantic warming and the first ever South Atlantic hurricane puts finis to the stand pat scenario. My particular theme of'gravitas', (in this case gravity) suggests that the moving concern in regard to 'Gravitic Flux' should allow us to mine both the liquid water and any excess ice which may threaten the stability of the ice cap. Antarctica will continue to change both under the 'natural' influence of the Milankovitch cycle, and the man-made influence of the ozone hole. But the mining of Antarctic ice and restive ice water with its consequent application to the Qatar depression, (just the biggest of four water-hungry depressions that could delay and possibly remove the imminence of the threat of rising oceans and contribute to the renewal of the natural preeminence of the Atlantic basin's evapo—transpiration cycle.) As nearly as I can see—this represents a plus plus for all areas of the globe! Please allow this idea to sink in for a bit. I will add the clinching carbon sequestration argument a little later in the manuscript.

But our 'Gravitas' author has not finished yet. Section 21, " what are the implications of our observations of the 19 97-1998 El Niño on prediction? Is ENSO more difficult to predict than we had thought?

The scientific community has made tremendous advances in forecasting El Niño in the past 15 years. For example, we had no forecasting capability at all prior to the 1982-83 El Niño. Many computer models correctly forecast that 1997 would be unusually warm in the tropical Pacific. That is a major advance by any measure, because just knowing that the tropical Pacific will be warm (or cold) a season or two in advance provides great leverage in making more reliable long-range weather forecasts around the globe. (This is a very optimistic message). On the other hand, the forecast models missed the rapid onset, the great magnitude, and the sudden demise of the 1997 98 El Niño, possibly due to weather noise that is inherently unpredictable more than about two weeks in advance. What that means is that there may be some inherent limits to how accurately we can hope to predict El Niño (admittedly a somewhat pessimistic message). However, in 1997-98 El Niño will serve as a stimulus for improving forecast models, because forecast skill is only limited by climate noise, but also by imperfect model physics, and incomplete and imperfect data for initializing forecasts. These are areas where we can certainly expect to see progress in the coming years, (again, an optimistic message)."

Question 22." **Was the strong 1998 99 La Niña related to severe winter weather in the Northern Hemisphere?**

> **"In 1998 1999 La Niña made itself felt in the US. The seasonal forecast for wintertime conditions, based in large part on the evolving temperatures in the tropical Pacific captured many of the large-scale patterns of temperature and precipitation of the continental US, from California to Florida. One forecast 'miss' was that the upper Midwest was predicted to be colder than normal this winter, but was a little warmer than normal, at least initially.**

Question 23." Why was El Niño such a big deal in 1998? . . . Three factors made reporting of the 1997-98 El Niño different . . .

1. the 1997-98 El Niño was the strongest on record, and it developed more rapidly than any El Niño of the past 40 years. As a result, we started to see its impact on weather, marine eco-systems and fisheries very quickly, and these impacts were spectacular. Early effects in August to October of 1997 included record flooding in Chile, marlin caught off the coast of Washington, the extensive smog cloud over Indonesia, and a quiet Atlantic hurricane season. The press is geared towards reporting sensational stories, and

this El Niño provided high drama through natural disasters and other unusual events.

2. In the past 15 years scientists developed new observational tools that allowed us to track the development of El Niño in greater detail than ever before. The new observations, from satellites and from sensors in the ocean itself, provided a day by day account of events as they unfolded in the tropical Pacific. These technological advances, providing high definition information on the tropical oceans and atmosphere system like never before, fueled a lot of interest in the press about El Niño, how we track it, and how it affects people's lives.

3. Another technological advance in the past 15 years was the development of long-range forecasting capabilities for predicting the evolution of El Niño sea surface temperatures, and the consequences of those temperatures on global weather. The effects of El Niño on North American climate are most pronounced in the winter season. Because the El Niño developed so rapidly, with record high sea surface temperatures in the equatorial Pacific by July 1997, forecasters could predict the full six months in advance with some reliability that the winter over the US would be very unusual. The credibility of these forecasts was high, because of the clearly identifiable impacts of El Niño earlier in the year, (see.1 above). The anticipation of an unusual winter motivated a lot of disaster preparedness efforts by local and state governments, the federal government, by businesses, and by individuals. This therefore caught the attention of the press. Once winter arrived, the predicted unusual weather set in, and that was also newsworthy. It turns out that the forecasts for heavy rains over the southern part of the US for the winter of 1997-98, and for an unusually mild winter in the Midwest proved to be largely correct. Record rains occurred in particular in California and Florida, two of the most populous states in the nation."

And finally, 24." has a reasonable, scientific body come up with any meaningful conclusions and/or predictions in the field of physical oceanography regarding the forecasting of the ocean

This somewhat lengthy discussion of El Nino is vital because of its intense relevance to the future of Southern California water choices and their consequent effect for Western United States and by extension Northern Hemispheric climate benevolence.

"Re: forecasting of the ocean, you may want to check out the following webpage: http://www.p mel.noaa. gov/tao/elnino/forecasts.html. Which summarizes a number of current El Niño/Southern oscillation (ENSO) forecasts. These forecasts rely on predicting tropical Pacific sea surface temperatures (SST) months to seasons in advance. Various kinds of forecast schemes have been developed. Some are based on the statistics of previous ENSO variations, whereas others are based on actually simulating future changes in ocean currents and subsurface thermal structure. ENSO forecasts are not perfect. However, they are sufficiently skillful at this point that individuals, corporations, municipalities, states, and national governments have used them to prepare for El Niño and La Niña events. We know that unusually warm or. tropical Pacific sea surface temperatures have major consequences for global climate and for Pacific Marine ecosystems. Forecasting Pacific SST's can therefore provide society with an opportunity to mitigate against adverse consequences or to take advantage of some of the positive aspects of ENSO related environmental change. The recent 1997—98 El Niño was the most recent example of success in ENSO forecasting . . ."

This extended slice of the present reality of ENSO

forecasting and preparation is intended to provide a compare and contrast with my ideas and speculations concerning Earth's climate and influences upon it. I have no complaints about taking 'advantage of some of the positive aspects of enso-related environmental change.

I will attempt now to remove some of the Star Wars fantasy from my own ideas.

For starters, the previously mentioned article from page 34 of Scientific American: Environment making a stand.

> "Three years ago in northern and central African nations that form the community of Sahel-Saharan states agreed to a continent wide belt of trees to combat the remorseless spread of the Sahara desert. This past June they laid the groundwork for the 'great green wall' of Africa by formally adopting a two-year, $3 million initial phase of the project. Green barriers against the Sahara have been around since the 1960s but most of them are small in scale. In contrast, the 'great green wall' will be 15 km wide and will involve stretches of trees from Mauritania in the West to Djibouti in the East—a distance of some 7000 km. The aim is to protect the Sahel Belt-the dry savanna south of the Sahara—and prevent its precious arable land from desertification. The trees would also provide a source of firewood, crops, and jobs. Projects to water these trees—say, by harvesting rain—could also help communities irrigate their fields all year long or even help them raise fish.

Pilot planting efforts, using local trees such as Acacia . . ., were scheduled to have begun in September. Funding for the entire project—perhaps its main stumbling block—still remains tentative. Signed Charles Q Choi." (Funding for the shelter belt is now above 116 million.)

It was not an accident that I underlined harvesting rain (above), nor was it an accident that I dwelt on the 'rain harvesting [Princess] of India'. The shortage of rainfall does something indelible to the consciousness of those who survive—look at the biblical Jacob and Joseph story. Also at Elijah and that 'cloud the size of a man's hand', and don't forget the farmer in James who 'waits patiently for the early and the latter rains. One spot in India is reputed to receive 82 feet per year! The Guinness 1976 book of world records cites three world record rainfalls: In 24 hours, 73.62 inches Cilaos, La reunion, Indian Ocean, March 15-16th 1952. In one month 366.14 inches, Cherrapunji, Assam, India, July, 1861. For 12 months 1041.78 inches, Cherrapunji, Assam, India August 1, 1860-July 31, 1861.

Rain harvesting applies to India, the Sahara, Southern California and perhaps to Arabia!

The potential for rain harvesting in India goes without saying: but I would like to call your attention to the culminating month of that world record year—30 feet in July at the peak of the solar gain! I've already called your attention to the absorptive excesses of rain darkened soils. But the line that offers renewed hope for the dry Sahara did you realize that water vapor is two thirds hydrogen—the lightest element. It's accelerated rate of rise (from rain darkened soils) acts as a vacuum pump *on the nearby ocean (or sea) bringing moisture laden air in to replace the rising column of heated vapor. All of this merely calls renewed attention to the science of reraining as best exemplified by the rain chart of 100 years in Nebraska. And soon to be re-proven by the $800 million 62,000,000,000 gallon reservoir in Florida. The purpose of that statement is to relieve the rainmaking professionals soon to be assigned to the Sahara of the seemingly eternal necessity* of repeating the rain making process for the *Sahara. Once started and vegetated, the process should have its own repetitive dynamic. One shouldn't expect that the rain, once begun, would achieve the constancy of Mount Waialeale, Hawaii, 5080 feet, 451 inches per year over a period of 52 years. 1920 to 1972 [about 350 days per year during which rain falls.] BUT WHEN WE FOCUS UPON THE MONSTROUS EXTREMES OF CHERRAPUNJI AND THEN CONSIDER THE ENDLESS YEARS OF ZERO RAINFALL IN OTHER PARTS OF THE GLOBE, IT BECOMES THINKABLE TO AIM FOR AN EASING OF THOSE EXTREMES!* (Guinness 1976 page 146.)

I will dip briefly into the forbiddingly hot political arena of Mideast politics, and then beat a strategic retreat to the (hopefully) friendlier shores of America. The Mideast's dire water straits and abundance of oil money sits strategically close to the overabundance of the Himalayan Mountains. It may only be a fortuitous accident of geologic history coupled with the play of present oil power politics. It's possible a deal for water could be struck as a plus, plus, for all participants if the timing of the use of that water on hungry soils was planned to draw onto the shore from the abundance of warm equatorial water in a siphoning action which contributes to a lessening of the annual excesses of " typhoons and hurricanes and cyclones."

Rain harvesting and re-raining theory confirmed in southern Missouri springtime

Perhaps I am repeating myself unnecessarily, but if water care strategically lubricates the land of the Mideast it can remove most of those previously mentioned hot dry dams (mountains) of impregnable dry air which contributes to the reversal of the normal tradewinds and thereby lead to the inevitability of El Niño. It *goes without saying that a greener Mideast would contribute sizably to the sequestering of carbon, thus partially fulfilling the terms of this contest! In keeping with my effort to present the colloquial side of the search for climatic wisdom—I have visited Table Rock Lake in southern Missouri during early October of the year, 2008. I discovered from the natives and permanent visitors to the area, Number one. The lake was higher that spring than at any time since the completion of the dam—late 40s to early 50s. I had heard of a series of multi-inch rainfalls coinciding almost precisely with the arrival of the spring sun and the greening of the plants. Number two. The natives confirmed my-(then tentative)—theory, that the albedo change (enhanced greenery and wet soil darkening) produced an almost mystical, magnetic, and irresistible attraction for further rainfall. Number three. My own conclusion: that these events offer an irresistible attraction to the desperately water short societies of the Sahel, and in the Mideast, and a rational hope that once begun—the heavenly cleansing waters will continue to flow thus justifying the initial extra effort to " prime the pump". For some Americans my extended concern for North Africa may seem somewhat surreal. It's time to shift to somewhat more familiar territory. This material from Science News February 23, 2008, pages 115 and 116 of that volume may help to bring it home.*

Southern California, fresh insights and recommendations

"Going down. Climate change, water use, threaten Lake Mead. If climate changes as expected and future water use goes unchecked, there is a 50% chance that Lake Mead—one of the South Western United States key reservoirs—will become functionally dry in the next couple of decades, a new study suggests.

. . . Lake Mead— the reservoir formed as the Colorado River collects behind Hoover dam, generates prodigious amounts of hydroelectric power. Over the past century, on average, about 18.5 km³ of water flowed into Lake Mead each year, says Tim P. Barnett, a climatologist at the Scripps institution of oceanography in La Jolla California.

Of that amount, about 2.1 km³ evaporate into the dry desert air or soak into the ground beneath the lake each year. What's left in the lake is more than spoken for: The amount of water drawn from Lake Mead this year to meet demand in cities as far-flung as Los Angeles and San Diego will exceed 16.6 km³.

And the situation will likely get worse, Barnett and colleague David W Pierce speculate in an upcoming water resources research (journal?). By 2030, the researchers note, annual demand for Lake Mead's water is projected to rise to 17.4 km³. Also, some climate studies suggest that the Colorado's flow will drop between 10 and 30% in the next 30 to 50 years. Using these data, as well as weather simulations that impose random but reasonable annual variations in river flow volume, Barnett and Pierce used a computer model to estimate the remaining useful life of the Lake Mead reservoir.

Thanks in part to the worst drought in the Southwest in the past 500 years (S. N: six/26/04, page 406), Lake Mead is now at about 50% capacity. If current allocations of water persist, there is a 50% chance that by 2023 Lake Mead won't provide water without pumping, and a 10% chance that it won't by 2013. Moreover, there is a 50% chance that Hoover dam won't be able to generate power by 2017, the researchers estimate.

'We were stunned at the magnitude of the problem and how fast it was coming at us,'says Barnett.

Results of the new study are 'fairly provocative, an eye—opener, 'says Connie Woodhouse, a climatologist at the University of Arizona in Tucson. Using estimates of river flow based on an average of the past century may be optimistic, she adds, because tree ring based reconstructions of the region's climate suggests that the 20th century was one of the wettest in the past 500 years. 'The more we learn about the Colorado River and its hydrology, the more worried we need to be,'says Peter H. Gleick, a hydrologist at the Pacific Institute in Oakland, California.—"

If that scare scenario is not bad enough, allow me to redouble the emphasis, The five year El Nino—1992—1998, —I know that adds up to six years, but would you rather have me say four El Nino's in a six year span. By the books, those El Nino's should have filled Lake Mead to overflowing but apparently they did not. This brings me to the crux of my argument! We are now going to expose the 'BIG LOSER'. And we are going to lay out enough options that the politicians can present something palatable for the voters.

Political options, El Nino and Lake Mead

First let me lay out the historical chart of Lake Mead Water Levels—Historical and Current dated 1/14/2010 from http://www. arachnoid.com/Natural Resources/ What you see there has to be interpreted by the chart maker; but there are a couple of missing teleconnections. The all-time high (1229) FT. suitably follows the date of the $13 billion damage attributed to El Nino. Was that water level attributed as an asset or a liability by the statistician? If you cannot yet draw a connection between El Nino and Lake Mead, look again at Lake Mead water levels from about '92 to '00. Again I have to wonder if the $40 to $100 billion damages figured positively or negatively into the height of water levels in Lake Mead. To me, that is a positive connection. It is an inescapable conclusion that Lake Mead needs El Nino! This means that Mexico, Arizona, and Southern California irrigators are dependent on El Nino and will be hurt if it is cured! The Chart and the kayaker's comments follow below:

Lake Mead Water Levels — Historical and Current

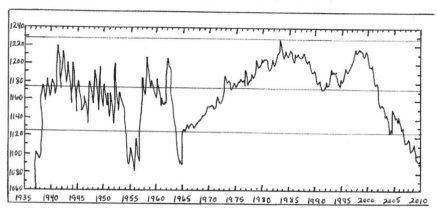

drawn from http://www.arachnoid.com/NaturalResources/

Amber A Leathers

The observant reader will notice a pattern of rapidly varying water height in the Lake Mead chart above from 1935 until the mid-1960s, after which the water level became more consistent in the short term. My theory is this smoothing was caused by the fact that Lake Powell, upstream from Lake Mead, began to fill in 1966, taking 17 years to fill completely (that would take us to 1983). It is reasonable to assume the people overseeing this filling operation took more water for Lake Powell at times of rapid flow, thus smoothing out the flow peaks and troughs that were seen in Lake Mead beforehand. Since that time I would guess that Lake Powell now absorbs the annual peaks and troughs once seen in the Lake Mead data, and acts as a buffer for Lake Mead. I would love to confirm this theory, but there seems not to be a convenient monthly water level database for Lake Powell, as there is for Lake Mead.

Update: In a 2008 report on the status of Lake Mead, scientists at the Scripps Institution of Oceanography predict there is a 50% probability that Lake Mead will be *completely dry* by 2021, because of climate change and unsustainable overuse of Colorado River water. The report concluded "Today, we are at or beyond the sustainable limit of the Colorado system. The alternative to reasoned solutions to this coming water crisis is a major societal and economic disruption in the desert southwest; something that will affect each of us living in the region".

—The scare line contained in the last paragraph—"there is a 50% probability that Lake Mead will be completely dry by 2021," . . . . lies near the heart of the American portion of this book.

The additional concern is mine but is derived from James Hansen's poignantly stated concern for the tipping point of climate possibly being initiated by Greenland's melting ice cap and burying the warmth of the Gulf Stream under the salt-free waters flowing from Greenland's water-lubricated icecap.

I am not as concerned in this particular direction now as I was at the start of this book. The changes already begun, Florida reservoir, less dumping of tanker residue, iron filings experiments; coupled with the attractive financial logic of timed interment and later extraction of Siberian methane lead to a reasoned hope for climate restoration.

My legitimate concern is coupled with the concept of teleconnections and the projected 7,000 km shelterbelt of trees planned and begun for the rehab of the Sahel. The incipient success of these projects will augment the indispensable trade winds whose death is the necessary precursor to the onset of El Nino.

Stated baldly, more trade winds equals less El Nino. Less El Nino equals a continued reduction in stored water for Lake Mead and the desert Southwest. Not enough time has passed in 2010 {now late February} to assess the effects of the spring El Nino, But http://www.gov/1c/region/g4000/24mo.pdf offers this depressing assessment. " Current runoff projections into Lake Powell are provided by the National Weather Service's Colorado river Forecast Center and are as follows: Observed unregulated inflow into Lake Powell for the month of January 2010 was .304 maf or 75% of the 30-year average. The forecast for February 2010 unregulated into Lake Powell is 0.325 maf or 77% of the 30-year average. The forecast for the 2010 April through July unregulated inflow is 5.8 maf or 73% of average." AND THAT'S FOR AN El Nino year!

It therefore becomes imperative to make rational plans for the water future of the desert southwest. I believe that it is possible to accomplish this goal within the ethical framework that I have stated. Restated, it is. We don't have the right to alter the world's climate except insofar as we attempt to repair damages that we may have done. But it also goes almost without saying that it is legitimate to anticipate and prevent catastrophic change.

The give-up side of the Lake Mead situation simply would alter the water contracts by noting that the contractual expectations were formed during the wettest period of the last 500 years. I don't believe that!

I believe that our expectations of normal were altered by human behavior some of it imperceptible and some of it glaringly obvious, over a period of 120 years. The easiest solution is to raise Charles Mallory Hatfield from the grave and pay his deserved wages. In addition to the physical impossibility, there is the painful political reality of being the politician who encouraged flooding. It's hard to know at this point which is the more difficult task. When we fully grasp the truth of reraining, it will become attractive to savor the January precipitation and expect its solar-augmented rising air to attract more of the richness of vapor hovering above the Eastern Pacific. The pang of temporary flooding can be alleviated by the certainty that almost all of January evaporation from Southern California will become deposits of useful snowpack in the Rockies. And so what if a little white stuff escapes over the Continental Divide to alleviate the Ogallala aquifer deficit! And it is entirely possible that the skills of modern rainmaking when applied near the California coast, when teamed with the Australian rain vats and scientifically selected deep-rooting trees may actually reduce the epidemic of mud slides that regularly capture nationwide interest and sympathy. But the reality of power politics must be fed on the hope that the return of an adequate snowpack for the Rockies will actually reduce the violence of the Santa Ana winds and minimize the Late Season Destructiveness of those forest fires.

For the hardened skeptics among my audience, and I count a good many of them as my best friends, please focus on the mystery that

I first saw on the History Channel captioned as the Mystery of the moving (uphill, southwest to northeast) rocks of Death Valley.

It's called the Agulhas current.

A sliding rock that has left a long track acthe surface of Racetrack Playa. Some tra
are hundreds of feet long! (See below f
several more sliding rock photos.) ©
iStockphoto / Steve Geer

What could make a 700 pound rock slide uphill? No one has witnessed it but their paths are plainly visible offering fertile soil for multiple theories. Let me add mine and attach it to the incredibly difficult feat of delivering moisture from the Pacific Ocean over the top of three mountain ranges to the dry bottom of Death Valley. That once a year or once in three years rainfall would be an obstacle for even the most patient observer. But the uphill sliding rocks offer concrete proof of the solar power that is harnessed by the color changes induced by moisture on the desert floor. The missing line in the theories so far

257

offered is this: When water vapor captures and holds the solar energy flowing into the Valley, its accelerating rate of rise leaves behind a vacuum that can only accelerate the already dessicatingly hot south winds that normally flow into Death Valley. Now, in case you haven't read the book, we step squarely into the dangerous realm of politics. The summary of the case against water. California can be perceived as the gatekeeper for the arid southwest. Remember the picture of the totally beclouded Eastern Pacific and the almost cloudless Western United States. In its most positive sense it can be viewed as the land of endless sunshine.[1] The needs of the photographic media [2] are best satisfied by hours of endless sunshine. The ripening fruits and nuts [3] are aided by more sunshine. Even the basements [4] of Southern California are protected by the barriers to ocean moisture that are heightened by the ever expanding roads, drainage ditches, and parking lots, [5] that hasten the departure of the occasional winter rains back to the relative anonymity of the ocean. The plaints, raised by the sliding mud hillsides conspire with the winter augmented and summer-dried shrubbery to raise in some minds an additional unwelcome mat for more water [6] Even the raucous noises of the 'Green tree-huggers' contribute to the continuing drought by denying oil drilling within 50 miles of shore, [7] {my own helpful interruption—If Israeli divers could cap and divert the salt springs in the Sea of Galilee around 1948, then it is not too much to expect oil drillers to cap and harvest the oil seepage around LA in return for the license to drill in 2010.} thus slowing if not halting the continuing natural oil seepage with its effective vapor barrier prolonging the aridity of the land. If that is not enough, three mountain ranges lie bestride Southern California naturally extending the many thousand year drought in Death Valley. This is just one side of the story designed to give skeptics a full reign to their choice.

If we look beyond this fearful array of political and natural forces prohibiting any but the uneasy and improper status quo; we can see another set of fearful options. The Northern California land owner's plaints have been mostly ignored or shelved for 100 years. The Yellowstone deprivation may have already slowed the natural teakettle regularity of Old Faithful. Any increase in drought conditions can only hasten the inevitable catastrophic explosion and will certainly handicap any efforts to harvest the mountain of geothermal energy beneath it. We've already discussed Lake Mead's plight in detail. And last but not least, the Ogallala aquifer has been receding for years. A substantial part of recharge can take place in the extensive Sandhills of Western Nebraska. The corollary expansion of carbon sequestration via the Bob Dole Plan can't happen until California lays out the welcome mat for more water. The proposed and planned remedies for Global Warming can only expand and heighten the civilized dryness of Southern California. The expanded rain storage cisterns anticipated for Southern California might be cleverly harnessed to extend and broaden the winter rainy season and thereby restore much of the Rockies snowpack. But it must be synchronized with extensive preparation for deep-rooted vegetation trees. [please allow me to insert a small natural exception to the tree argument which has been made successfully in many times and places. This strange but welcome line comes from the book; Prescription for Nutritional Healing, 2nd edition James F Balch MD. Lsbn 0-89529-727-2 copyright 1997 by Phyllis A Balch. From page 49 "Alfalfa One of the most mineral-rich foods known, alfalfa has roots that grow as much as 130 feet into the earth . . . good for fasting . . . It contains calcium, magnesium, phosphorous, potassium, plus all known vitamins . . ." It's no wonder that the salient truth escapes when this wondrous food is discussed. The "roots can grow as much as 130 feet into the earth." When this truth is placed on top of the sliding hillsides of saturated soil and can be harvested in season down to the fire-retarding two-inch level, It becomes a powerful rebuke to the home-owner who naturally resists any brushy growth that adds fuel to the late season Santa Ana winds and forest fires. The only thing missing from this powerful argument is the length of time for the roots to grow that deep and experimental evidence that it will be deep enough to work. All of this argument places me squarely and ethically on the side of rewelcoming ocean rainwater on land. And, of course, I'm already on record as recommending the hydropower "river" into Death Valley. This is the do-something side of the skeptics choice and in its caricature-like simplicity it by-passes all of the agonizing niceties of the timing of actions. As a considered blessing to a portion of the movie industry: One could cede or sell the southwest (dry) corner of Death Valley to the moviemaker's Exodus from the fleshpots of Los Angeles. There they could fully embrace the role of ombudsman to the new wetter order. And now back to my often reiterated but still relevant question. How much less ozone is created on the rising columns of Amazon rain forest oxygenated air? And would it be restored in Western United States if we give rain water a chance to fulfill Sprengel's Law?

Chapter14

Raising our sights to the big picture

And now I am prepared to join James Hansen in his concern for the tipping point of climate. This is not uniquely my bias. All of the Northern Hemisphere should join in the concerns noted below. But my concern (bias) is the proximity of Nebraska to the major source of climate tipping. Yellowstone. But the discussion of water options for Southern California may point to some somewhat reliable buffers for the disaster to come. The picture of the Sliding Rocks of Racetrack Playa display a southwest to Northeast push that aims like a sniper's rifle to Yellowstone. Any water introduced to Death Valley could enable the evaporation that could resupply the giant boiling teakettle, which is Yellowstone's Old Faithful, and possibly contribute to the minimization of any future disaster. My gut tells me that the mystery of the uphill moving rocks in Death Valley cannot be solved without reckoning a near total conversion of the solar spectrum to rising vapor which then promotes the returning rush of surface replacement air at floor level in Death Valley.

The sketchy option of harvesting Yellowstone's geothermal receives an added push toward reality by the addition of natural (non-salinated) moisture available to the electrical hydrological, and metallurgical engineers in charge of the water to steam portion of the project. These few paragraphs hinting at a possible Death Valley to Yellowstone water connection do not provide enough motivation for any California politician to begin a multi-billion $ project. Let's supply some facts. This history of Old Faithful is taken from the Internet with documentation intact. My interpretation is as follows: The 1871 account describes spouting "at regular intervals nine times during our stay, the columns of boiling water being thrown from ninety to one hundred and twent-five feet at each discharge, which lasted from *fifteen to twenty minutes.* We gave it the name of *"Old Faithful."* The interpretation of the decreasing eruption and increasing interval which I find lacking is repeated below: "Over the years, the length of the interval has increased, which may be the result of **earthquakes** affecting subterranean water levels. These disruptions have made the earlier mathematical relationship inaccurate, but have in fact made Old Faithful more predictable. With a margin of error of 10 minutes, Old Faithful will erupt 65 minutes after an eruption lasting less than 2.5 minutes or 91 minutes after an eruption lasting more than 2.5 minutes. The reliability of Old Faithful can be attributed to the fact that it is not connected to any other thermal features of the **Upper Geyser Basin. "My interpretation which you might be familiar with by my extensive discussion of Yellowstone and Southern California water harvesting suggests that the shortening of the Old Faithful eruption cycle is man (and nature) caused. In short, Mother Nature's safety valve—the 15 or 20 minute long eruptions in 1871 were threatened by the gov'ts tampering with the elk population and the consequent disappearance of the beaver ponds. The not-so-obvious harvesting of Northern California waters by L.A allowed the drier land surface to heat up and become periodically repulsive to normal rainfall. That there was natural oil seepage around L.A. is well-known . . . That poor drilling practices around L.A may have contributed to the righteous** anger of Northern Californians and by extension even Yellowstone is a little known truth that may require getting used to. It now should not require much additional begging for California to put out the welcome mat for more water for their own interests and claim with some truth that they are acting for the benefit of the whole nation.

Extending the Growing Season

Careful readers should be reminding me at this point that I can't finish this book without the 'extensive plan to sequester carbon' on the Great Plains. One answer is that I already have. The Norwegian traffic pattern that preserves motion by minimizing T-bone collisions is nearly perfect for adapting to the Plains phenomena of center pivot irrigation systems. Before I elaborate on the niceties of that change, I need to remind Californians that their welcome to the increased snowpack on the Rockies is indispensable for the Minimization of the still-expanding dead zone in the Gulf of Mexico. In simplest terms No snowpack no Spring Flood of the Mississippi-Missouri No spring flood= no sediment leaving the mouth of the Mississippi, No sediment= no annual interment of algae produced by the excess of nutrients pouring off of the plains. It begins to appear that as California goes—so goes the nation.

Let's take a bit of that guilt from the Californians! The appropriate modification of Nebraska intersections to enhance the merging speed of traffic, leaves a diamond shape or interstate cloverleaf to preserve traffic speed while not threatening the enhanced crop circle. The water advisors for the State of Nebraska are already on record as advising water retention basins at the corners of each pivot system. At a very minimum it gives a visible yardstick to any excesses of fertilizer application. My water-mining friends in Nebr. [50+ wells at last count] assure me that the only water lost is when natural rainfall occurs during active sprinkling. The additional portion of logic to that budding pattern is supplied from the internet: *http://alcoholcanbeagas.org/*

site/bookmenu/360 The speaker . . ."asked the crowd if anybody knew of any good feedstocks other than corn, and of course, I piped up with my cattail rant. He started laughing—that was his next topic. And they've been doing a lot more work with cattails that I've heard about, it's far, far better than I thought. You don't have to dig up the roots, the bottom 3 feet of the tops are just as good. And you can get much higher yields both of starch content (up to 70%) and tonnage, they've gotten 70 tons per acre. [200 bu. Corn yields less than 6 tons] He proposes changing all sewage treatment plants to growing cattails for ethanol. And making ethanol out of it is just as easy as any other starch. Blum says it's by far the best crop for ethanol where you can't grow sugarcane, and maybe even for there too. Totally sustainable, permaculture, etc. What he was also big on was returning the spent mash to the fields, even with corn—in fact has been issued a patent on using the spent corn mash as both a herbicide and fertilizer—natural 'weed and feed'. He patented it just so Monsanto et al, couldn't, and gives permission . . . ." There is more to the remediation side including sequestration of methyl mercury but the fact that it is recommended for Sewage Treatment plants is a strong enough recommendation for me. If all center pivot sections were converted to the traffic augmenting diamond shape and the interior of the diamond became a wetland reservoir, the intersection could then become the most profitable portion of the farmer's field and the ease of harvesting wouldn't harm its attractiveness either. (The picture below is the redrawn intersection sample.)

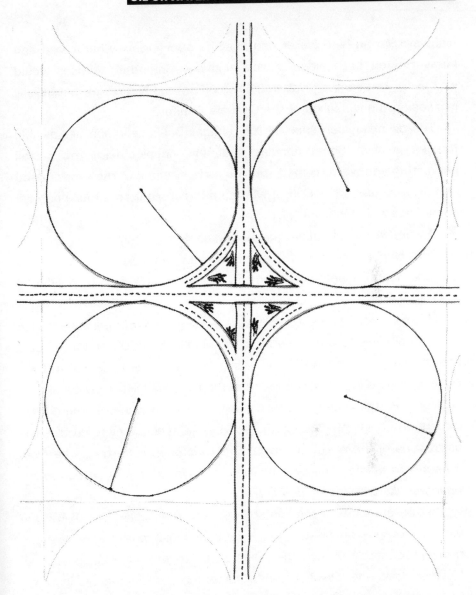

Keep in mind that though the graph above is a good concept of the Ogallalla Aquifer recharge basins process, the wetlands size, should encompass the arch from road to road.

My MODOT critiquer reminded me of this feature, of i80 across Nebraska. Where i80 runs beside the Platte River. The ditches were engineered to serve as wetlands reservoir by laying a pipe under the highway. Spring high water can best be appropriately harvested and treasured on both sides of the highway. Something similar would well serve the center pivot overflows all across Nebraska, potentially increasing the recharge of the Ogallala Aquifer.

The clinching sentence to this almost Sci-Fi rendition of possible futures is this: Some flooding sediment coupled with the cattail harvesting and water remediation could bury some of the excess dead and dying algae in the Gulf of Mexico and return it to a more natural abundance of life in the Gulf.

A brief review of salient (largely un-dealt with) facts.

The hidden portion of the global warming hysteria is that the micronutrients made available to the skies from 1935 thru' WWII and even extensive

Nuclear testing somehow **may have** made a micronutrient contribution that actually leveled the so called inevitable climb in CO2 and may have. actually declined a bit in the two-year period from 1962 to 1964. Even without laying out and dissecting all of the reasons for the period of CO2 decline, I am convinced that we can curb the growth of airborne carbon without draconian changes to the activity of the economy or curbing the aspirations of the world's peoples. *But that does not exempt us from a deep and abiding concern for the toxic byproducts of carbon in the hands of the wizards of chemistry.*

Astute students of the climate will note that I have separated the big word evapo-transpiration into its manageable constituents. We can now appropriately phrase the question as to the super-abundance of Northern cabbages. Was it the sunshine, the dust augmented increase to the rainfall cycle or the micro-nutrients in the dust? Or better still a timely combination of all three. We must not fail to note that any extra water vapor in the air becomes an ultraviolet magnet, (perhaps vacuum cleaner is a better description). Who knows? Perhaps continental plants respond to the absence of ultra-violet as enthusiastically as sea surface flora and fauna (Read Phytoplankton).

At any rate, like a suburban gardener we need to pay increasing attention to the need for micro-nutrients at any point that we desire greenery to prosper.

It is definitely feasible to declare that at least for Post WW2 that there was enough CO_2 and micro-nutrients available north of the Arctic Circle to trigger extensive growth. But let's come closer to home–now—from Chris Horner-

Politically Incorrect . . .—p. 75 "Greedy North Americans Using More CO_2 Than We're Producing," "I will concede, the measurements are not perfect, they are just pretty good; they come from Science magazine and they are the best numbers currently at hand–America today is apparently sinking more carbon out of the air than it is emitting into it. What's doing the sinking? In large part, re-growth of forests on land that is no longer farmed or logged, together with faster growth of existing plants and forests, which are fertilized by nitrogen oxides and carbon dioxide 'pollutants.'" Peter Huber, "Hard Green: Saving the Environment from the Environmentalists

Almost the same thing was stated with admirable brevity in 1828 by Carl Sprengel. From Wikipedia, http://en.wikipedia.org/wiki/Timeline_of_environmental_events "Formulates the Law of the Minimum stating that growth is limited not by the total of resources available, but by the scarcest resource." *And the scarcest resource for 1/3 of the land surface of the globe is water!*

I am personally curious about any positive changes that may have occurred in the phytoplanktonic world following the 1966-80 period? **That curiosity cannot *rise to the level of my interest in plankton for the* period 1962-1966 but I am unaware of anyone's interest in plankton at that time. My curiosity is—best focused by Sebastian Junger's book—the "PERFECT STORM" p153 &4 "There is some evidence that *average wave heights are slowly rising, and that freak waves of 80 or 90 feet are becoming more common. Wave heights off the coast of England have risen an average of 25% over the past couple of decades, which converts to a twenty-foot increase in the highest waves over the*** next half-century. One cause may be the tightening of environmental laws, which has reduced the amount of oil flushed into the oceans by oil tankers. Oil spreads across water in a film several molecules thick and inhibits the generation of capillary waves, which in turn prevent the wind from getting a 'grip' on the sea. PLANKTON RELEASES A CHEMICAL THAT HAS THE SAME EFFECT, AND PLANKTON LEVELS IN THE North Atlantic have dropped dramatically. Another explanation is that the recent warming trend—(book pub. In 1997)—some call it the greenhouse effect has made storms more frequent and severe. Waves have destroyed docks and buildings in Newfoundland, for example, that haven't been damaged for decades . . ."

A similarity of function does not imply a causal relationship! The evidence of 1962-66 suggests rather strongly to me that a certain kind of particulate matter (acid-reducing) holds the key for the necessary renewal of ocean phytoplankton.

Or for that matter has any scholar taken note of vegetation changes within the Nuclear Fallout Period? Approximately 1945-1980? What has appeared in print–Chernobyl, Three Mile Island, even irradiation of supermarket vegetables seems minimalist and inadequate to me. Even the extensive and costly remediation efforts aimed at Bikini Atoll show relatively little concern for a durative study of plant life within the lagoon!

But I am less excited about that question than I am concerned as well as excited about applying Sprengel's logic to the current fad of sprinkling iron filings on the ocean to reinitiate the algal and phytoplankton bloom–because in my humble estimation–the What as well as the Where of seeding as well as the When of the ocean seeding holds great promise as well as great responsibilities for the shore-ward recipients of the resulting evapo-transpiration. The most profound reason for aiming shoreward the increased evaporation resulting from the sea-born phytoplanktonic murkiness are the 200 foot long strands of algae just offshore of the Columbia river. Understanding the scarcest nutrient in this area of bountiful growth could only enhance our understanding *of the need for intimate continental-oceanic interaction.* I have already mentioned a stopgap alternative to the expected (dreaded) decline in Amazon (transpiration aiding) output–but iron filings at that point would be inadequate for several reasons; number one because they don't begin at the logical beginning of the process. The logical place to start should be on the north shore of Australia where the Gulf Stream is reputed to have its genesis; but the major sin was not committed there (to my knowledge) so I am here going to abbreviate my exaggerated search for first causes and focus on second causes–the seasonally abbreviated phytoplankton bloom around the Antarctic Ice Cap.

Since I don't permit myself unlimited recourse for change–only redress of grievances I will focus once again on the ozone hole and the potential consequences of a somewhat belated healing of that vegetative crippling ultraviolet (and cancer-causing) invasion.

I believe that I have already identified my choice for restoring normal phytoplankton bloom. Quit dumping used oil into the ocean and rinse the tankers into one of the five overheated depressed evaporative zones onto the three continents already blessed with space and science to remediate the oil into useful and beautiful surface based greenery. The motives for restoration (that is pre ozone hole behavior). Are adequately revealed in a cautionary exploration of some present-day side-effects. From http://web3.infotrac.galegroup.com/itw/infomark/601/113/87496777w3/purl=rc1_ITOF_0_A115495 . . . 2/9/2005 p. 3 of 12

"Some of the consequences of accelerating CO2 buildup, such as melting polar ice and damage to forests, are well known. Others are relatively obscure, but no less devastating." (*my vision of reality waxes a trifle indignant at this point, because I tend to value CO2 for its beneficial fertilizing effect, but I will content myself by substituting the term oceanic sealing,) The above material sidesteps my personal bias toward saving and harvesting Yellowstone, But I truly believe that it fulfills the prevention of the 'Tipping point of Climate' that has been so eloquently described by James Hansen.*

Again, we're drawn back to the ancient record. Russell Graham, chief curator of the Denver Museum of Nature and Science, has noted at least 63 sudden climatic changes in the last 1.6 million years, an average of one every 2,000 years. As Gregg Easterbrook noted in 'A Skeptical Guide to Doomsday,' a 2003 Wired article, 'Ten thousand years have passed since the current pleasantly temperate period began so another sudden shift is overdue. The notion that greenhouses could trigger such a rapid change keeps serious scientists up at night . . . .

And since scientists today have little understanding of past climate flips, it's impossible to say when the next one will start."

He doesn't say whether he gets his evidence from one or more of Greenland (or elsewhere) ice cores; It certainly contrasts with the Milankovitch cycles but, it does reinforce the evidence from undigested food in mammoth's stomachs i.e. "sudden climate change with little forewarning for the big hairy beasts."

"What may have killed mammoths and other large mammals 11,000 years ago is a glacial melt in Canada and the northern U.S. that led to extreme winters and summers, and also to a more homogenous landscape of forests and grasslands. Complex combinations of plants disappeared"

It is now time to offer a little common sense fillip to the stated confusion and lack of understanding of past climate flips–here it is. The only way that the atmosphere can deliver (an ice age worth of cold) enough is because the Ocean(s) in one of two forms (volcanic + or vegetative-) stores and releases enough heated moist air to form that impenetrable barrier to incoming radiation (both as clouds and later as snow or ice on earth's surface.) The once formidable whales as surface harvesters have been superseded by oil spills, nets, and don't forget the eastern and western Pacific Garbage Dumps which contribute their share to the altered chemistry—even my Father's soil retaining conservation efforts may have contributed in a minuscule way to the exposure of New Orleans to Katrina. One thing is certain–the runoff from my Father's farm in the forties posed a significantly lesser impediment to Ocean Biota than the Anti-biota (ic) that spills from our American Pharmaceutical Establishment.

Once again wrestling with the Ozone Hole(s)

One of the biggest culprits in the vegetative minus department is the Ozone Hole (s) The Ozone Hole coupled with reduced evaporation has contributed to a dryer climate around the equator, but the long term effect of increased ocean permeability has probably already contributed to the larger storms that are central to our fears. Those downpours are the opposite of the soil softening drizzles that contribute to long term vegetation and root health. At the risk of being a "Johnny One Note" I reiterate by Father's mantra. "It's not as if we had that much less rain in the dirty 30's It's that the rain that did come, fell onto sun baked soil and ran off."

The prescription for the repair of the Antarctic Ozone Hole is all but spoken aloud—from the internet http://science.nasa.gov/headlines/y2001ast11oct_1.htm The article titled, "Planetary Long Waves Drive Polar Temperatures and Affect Ozone dangerous beauty."

Polar Stratospheric Clouds (PSCs) are common in Antarctica, but a rare sight in the Arctic. They form when temperatures in the stratosphere become extremely cold—below-78 degrees C. PSCs spell trouble for ozone; tiny ice crystals and droplets within the clouds provide surfaces where CFCs are converted into ozone-destroying molecules. Credit: Lamont Poole, NASA.

Spring is the season when sunlight can trigger the chemistry of stratospheric ozone destruction. But Earth's two poles react differently to the coming of spring. Springtime in Antarctica heralds a large ozone hole, while springtime in the Arctic (six months later) often brings above-average ozone concentrations. Global warming could alter this familiar pattern, though, by chilling the northern stratosphere and producing an ozone hole there as well."

With these two paragraphs in front of me its easy to prescribe a fix for the Ozone Hole. But before I give you the short sentences let me quote once more from the document and perhaps we'll understand why they couldn't make the connection. "It so happens that stratospheric cooling can be a curious result of global warming. Greenhouse gases, which trap the heat radiating from Earth's surface in the lowest layer of the atmosphere, reduce the heat that reaches the stratosphere. In effect, greenhouse gases cool the stratosphere by insulating it from the warmer Earth below." *My version is much simpler; Stratospheric dust makes for a much more direct route for upper level heating and obviates the effect of greenhouse gases at whatever level. For Antarctica a blanket of soot and a duke's mixture of calcium and sodium applied to the PSC's in a timely (springtime) fashion should minimize the ozone hole.*

The reasons for doing it are much more than the simple "you broke it–you fix it."

Although there is plenty of that blame to go around, auto air-conditioning, home AC, wrecks and simple aging, along with its handy use as propellant in hair spray,–. makes for many opportunities for finger pointing and blame–mostly aimed at the hyper-active apex–America. Simple blame-casting won't get us anywhere let's go deeper.

From the internet a summary by Virginia Morell of a Smithsonian magazine article of February 2007. Titled "Ahead in the Clouds " Susan Solomon helped patch the ozone hole." Now, as a leader of a major United Nations report–out this month–she's going after global warming. A no-nonsense 51-year-old atmospheric chemist, she's co-leader of the massive new study, along with Qin Dahe, a climatologist from the China Meteorological Administration in Beijing, Solomon will become the public face of the U.N . . . report, in charge of presenting the best scientific thinking on the subject of global warming and the evidence that it is caused by the burning of fossil fuels. 'The science is strong,' she says, 'and we'll be presenting a consensus view.'

To reach that consensus, Solomon logged more than 400,000 air miles over the past four years and held dozens of meeting with the reports of more than 500 authors. 'This much I can say: the climate is changing and quite noticeably,' she says shortly before the report is released . . . she suggests that policy makers (and the rest of us) have reached a critical moment in our dealings with or failings to deal with, climate change: 'The effects will vary from region to region, and the challenge that society will face is to get people to think beyond their own back yards and to make judgments about the risks they're willing to take.'

Maybe as the climate continues to warm, the ice caps won't melt; maybe a rising sea level will be offset by some other unforeseen event. She's reminded of the scene in Dirty Harry, in which the cop played by Clint Eastwood confronts a criminal: "You've got to ask yourself one question: Do I feel lucky? Well, do ya' punk?" Solomon says, 'That's what we as a society have to decide. Will we . . . make some changes in our behaviors? . . . Scientists who study climate change or other environmental problems often feel like Cassandra, the mythical prophetess of the Trojan War who was cursed to foretell the future but not to be believed. But Solomon doesn't see herself that way at all. She's more optimistic than many of her climatology peers, and with good reason: she was a prime mover behind one of the most important global environmental turnarounds in history, repairing the hole in the ozone layer above Antarctica.

August, 1986 . . . Although she had no field experience, she'd been chosen leader of the first National Ozone Expedition to the Antarctic in part because she was one of the few scientists to take the ozone hole seriously. And she had devised a theory to explain it . . . CFC's . . .—they drift with winds up into the stratosphere. There, ultraviolet rays kick chlorine atoms out of CFC's and each freed chlorine atom goes on to break apart as many as 100,000 ozone molecules.

The scientists who demonstrated the ozone-eating ability of CFC's–they would later earn a Nobel Prize in Chemistry for the work–believed that the degradation high in the atmosphere would be slow, taking perhaps hundreds of years. Yet the ozone shield was disintegrating quickly. 'And the thinning wasn't supposed to show up over Antarctica first,' Solomon recalls, 'since that's not where these chemicals were being used.'

She came up with an explanation after noticing something that others had overlooked: the ozone depletion was occurring in the lower stretches of the ozone layer, about eight miles above the earth, rather than in the stratosphere's upper reaches. A self-confessed 'weather-weenie' who loves watching summer afternoon thunderheads over the Rockies, Solomon knew that unusual clouds often form above the Antarctic each Austral (southern) winter and spring. They are so full of ice crystals they shimmer like abalone shells. The crystals contain nitric and sulfuric acids, and Solomon speculated that when chlorine compounds came into contact with these cold, Acidic clouds, the Chlorine was transformed into rabid ozone-eating chemicals.

Today more than 180 countries, including the United States, have signed the 'Montreal Protocol on Substances that Deplete the Ozone Layer' as well as amendments calling for a complete ban on CFCs. In 2,000, President Clinton awarded Solomon the National Medal of Science for her ozone hole research; the citation praised her 'for exemplary service to worldwide public policy decisions and to the American public.' She was one of the youngest members to be elected to the National Academy of Sciences, the country's most elite scientific organization . . . Climate scientists predict that the ozone hole over Antarctica will disappear by the end of this century and the ozone layer over the rest of the planet will thicken back up. 'With luck, I'll live long enough to see the layer close to being fully restored,' Solomon says. Part of that environmental success can be credited to the company that manufactured most of the world's CFCs, Dupont. It announced in 1988 that it would voluntarily stop CFC production, and company chemists soon devised replacement chemicals for air conditioners and refrigerators.

Solomon knows the current climate crisis won't be as easy to solve as the ozone problem. (THE OZONE HOLE DOES NOT CONTRIBUTE NOTICEABLY TO GLOBAL WARMING.) . . . ."—

I have to stop this extended laudatory article on Susan and offer my own praise to her accomplishments but I have to disagree vehemently with two parts of the article. It's wrong to conclude that the problem has been solved. She has identified the problem and possibly stopped the rate of increase of the ozone crisis but in 2006 the Antarctic ozone hole reached record size. While that may possibly be attributed to the increasing precision and sophistication of measurements,—I think not. Here follows two dramatic proofs that the ozone hole contributes spectacularly to global warming: First from the Kansas City Star, Monday Jan 4, 08 from A 4.

"West Antarctica losing more ice Warmer ocean current is suspected of speeding up losses in area where loss had been slight. (The Washington Post) Washington } Climatic changes appear to be destabilizing vast ice sheets of western Antarctica that had previously seemed relatively protected from global warming, researchers reported Sunday.

"While the overall loss is a tiny fraction of the miles-deep ice that covers much of Antarctica, scientists said the new finding is important because the continent holds about 90 % of Earth's ice, and until now, large-scale ice loss there had been limited to the peninsula that juts out toward the tip of South America. In addition, researchers found that the rate of ice loss in the affected areas has accelerated in the last ten years–as it has on most glaciers and ice sheets around the world.

'Without doubt, Antarctica as a whole is now losing ice yearly, and each year it's losing more,' said Eric Rignot lead author of a paper published online in the journal Nature Geoscience. The Antarctic ice sheet is shrinking despite land temperatures for the continent remaining essentially unchanged, except for the fast-warming peninsula.

+-The cause Rignot said, may be changes in the flow of the warmer water of the Antarctic Circumpolar Current, which circles much of the continent. Because of changed wind patterns and (less-well-understood) dynamics of the submerged current, its water is coming closer to land in some sectors and melting the edges of glaciers deep underwater.

Rignot said the tonnage of yearly ice loss in Antarctica is approaching that of Greenland, where ice sheets are known to be melting rapidly in some parts and where ancient glaciers have been in retreat. He said the change in Antarctica could become considerably more dramatic because the continent's western shelf, an expanse of ice and snow roughly the size of Texas, is largely below sea level and has broad and flat expanses of ice that could move quickly."

I'm going to leave the 'changed wind patterns' alone temporarily and focus on the (less-well-understood) portion. If you have focused on my arguments at all, the ultraviolet invasion (destruction or inhibition of surface phytoplankton) and its consequent reduction in surface murkiness (reduced biota) allows for deeper penetration of radiation and then the subsurface attack on the continent's western shelf (by warmer waters) is no mystery at all and is directly linked to the ozone hole. The fact that the Southern Hemisphere is now experiencing enhanced springtime closeness to the sun (courtesy of the Milankovitch cycles) places this observation squarely in the middle of the raging debate: "Is it natural or is it manmade?"

I can contribute a sentence or two (from pre-history—i.e. before 1945) to the debate.

In the extensive debate that surrounded the 'Dust Bowl—causes and cures,' it was more or less a given that ocean biota was on a 22 year ebb and swell cycle that coincided with sunspot activity. In Susan Solomon's short bio sketch there is no reference to 'natural' ebb and swell of ultraviolet impingement or of 'Natural' cycles of biota. I'm convinced that this debate should continue. Whatever shreds of ethics remain in our national treasury could be invoked to produce compensatory behavior.

First South Atlantic Hurricane Documented

To continue with the case against *(the* **singularity of the Antarctic)** ozone hole. (Do I need to?) This from USA. TODAY.com Posted 3/26/2004 by Jack Williams, "The first hurricane ever known to have formed over the South Atlantic Ocean could threaten the coast of Brazil by early next week.

Satellite images 'give all the appearance of a hurricane' with sustained winds faster than 75 mph, says Jack Beven of the U. S. National Hurricane Center in Miami . . .

He and other hurricane specialists at the Center are helping Brazil's civilian and military forecasters track and predict the storm, something Brazilian meteorologists have never had to do.

'We are trying to run our hurricane forecast models, but they are structured for the North Atlantic,' Beven says. 'Some just flat out refused to run.' The storm has no name since unlike in the parts of the world where hurricanes and similar storms are common, forecasters have never made a list of South Atlantic names.

Beven says meteorologists can't be sure that a hurricane never formed in the South Atlantic before satellites began keeping an eye on all of the world's oceans in 1960. In fact, Beven said, if a hurricane had formed before 1960 where the new one is now located, no one would have known about it.

Meteorologists, know of two South Atlantic storms that probably reached 39 mph—tropical storm strength. The latest—Jan. 19 (04). A storm that formed off the coast of Africa in April 1991 is also believed to have reached tropical storm strength, but like the January storm, did not come close to growing into a 75 mph hurricane.

While the tropical part of the South Atlantic Ocean has large areas with the 80 degree or warmer ocean temperatures needed to sustain hurricanes, upper atmosphere winds are usually blowing much faster than winds near the surface or in the opposite direction. Such 'wind shear' can rip apart storms before they grow into hurricanes."

The reported aftermath—from theage.com.au–Sept 26, 2005—"Hurricane Catarina barely rated a mention in the Australian media when it pummelled about 20 towns on the southern Brazilian coast in March, killing at least three people and flattening hundreds of homes.

Meteorologists, though, were stunned. The arrival of winds of more than 140 km/h over Santa Catarina state shattered a scientific consensus that it was impossible for a hurricane to form in the southern Atlantic Ocean. Accepted wisdom had previously said its waters were too cool and the difference between wind speeds of its upper and lower atmosphere too great." Do I need to remind you that the older charts of ocean temperatures show no portion of the South Atlantic–exceeding 74 degrees. Maybe this is simply a case of more precise measurement? But at the risk of being labeled a 'rumor monger' I'll supply another little factoid from my reading which I will not bother to document because I still do not know how to interpret it. Some nameless scholar has announced that the ebb and flow of the Milankovitch cycle produces a 31% increase in solar horsepower when the eccentric path of the earth's orbit is mated with the extremes of tilt–and that union is now being consummated at the South Pole during its necessarily foreshortened spring and summer (Sept through March) I only include this data (opinion?)—to contrast with the business as usual tone of Virginia Morell's article in Science.

New section—Healing the Ozone hole I had to stop The preceding data makes me almost physically sick. Not the facts. That southern hemisphere deep waters are heating under the increased onslaught of ultraviolet aided sunlight is no big surprise. The shock comes from Solomon's name and deserved prestige being attached to two opinions. 1st "Susan Solomon helped patch the ozone hole."

2nd. "The ozone hole does not contribute noticeably to Global Warming." While the 2nd statement is technically true, the advent of South Atlantic hurricane (s) and the warming erosion of the 'Texas-sized' West Antarctic ice field makes reality obvious even to a novice.

That 'unnoticeable' warming is now front page news and by her own admission will continue for the next 50 and maybe 100 years. If her opinions are accepted as gospel by the legion of

Climatologists, they will study the side effects of Global Warming secure in the knowledge that the problem will provide job security for a lifetime. *Freon is not the whole problem—the lack of water vapor* and clouds is the problem and it's related to the doubled width of the Sahara in the last fifty years as well as the spiking of the oceans with contaminated oil. Overheated land and oily oceans do not make for an attractive marriage.

Teleconnections: from the Eastern Pacific Garbage Dump to soil retention in the Sahara-Sahel.

And Finally the long awaited introduction of the 'young lions of NASA by the former head of Oak Ridge National Laboratory, Don Trauger— from his book HORSE POWER TO NUCLEAR POWER PAGE 423. "We have a heavy responsibility to future generations as we continue to populate the earth. I have read with wonder and amusement of plans to move people to other planets, the moon, and asteroids, even to mine those sources for materials, water, and perhaps energy. No doubt that will someday be feasible, but it surely is easier to populate the Sahara desert, which is well supplied with air and heat and is underlain with some water and oil. Until we see the Sahara widely utilized, space ventures as solutions to earthly problems other than in some areas of research must be viewed with skepticism. We occupy a wonderful planet that struggles to support the life and luxuries that many now enjoy; let us put emphasis on preserving it responsibly and forever for the benefit of all people."

The return of water to the Sahel (and Sahara) is in many respects a natural and somewhat overdue process—Perhaps the help that Nasa can best offer is a perception of best times for humans to catalyze the return.

There are far more serious problems related to 'patching the ozone hole' than the job definitions for climatologists. Those problems can only grow the longer the solution is post—poned. When the phytoplankton is allowed to return to 'normal,' the temporary addition of surface 'murkiness' and solar gain will combine with the latent deep ocean heat to contribute to some initially spectacular increases in storminess.

I can't predict the location and timing of the storms but I choose to regard the problem as opportunity and I also choose a preferred site for the atmospheric largesse that is the guaranteed result of prolonged ocean elevated temperature. That site is the Sahara desert; specifically the Qattara Depression west of Cairo, Egypt. (There are already a number of experts at work on repairing the Sahara and they can only be aided by my desire to harvest the Eastern and Western Pacific 'Garbage Patches' and convert that waste plastic into valuable tarmac to form a water-holding seedbed for 'wearing of the green' for the Sahara.) On the other hand Australia seems to be adapting at its own pace to the enhanced irregularity of Antarctic-triggered storminess–260 gal. catchment basins attached to the eaves of their houses–. Let well-enough alone. But consider this data From Godlike Productions www.godlikeproductions.com–date Aug 11 2005. "It snows on the south-east of Australia . . .

Snow fell Thursday for the first time since 1986 in Melbourne, the capital of the state of Victoria and with Hobart, in Tasmanie . . . Usually, it snows only on the mountainous areas of the Galle News of the South, Victoria and Tasmanie . . . The cold succeeds an autumn softer than normal, and in particular the hottest April ever recorded in Australia, specified Scott Williams . . ."*This material begins to reinforce Robert W. Felix's theme that 'OCEAN WARMING BRINGS THE ICE AGE.' AND tentatively contradicts the heartfelt relief of the Global Warming skeptics when surrounded by cooler than normal temperatures and abundant wintertime precipitation*

A Brief Cooling dip into fantasy land. Piri Reis and Antarctica

Of the potential negatives one that probably belongs in the science-fiction realm deserves mention as well as hoots of derision. Ancient (Piri Reis)? Turkish maps (15th century) show a topographically correct ice-free map of Antarctica. These maps could not be confirmed by modern techniques until (approx. 60's) satellite radar. The only theory that could account for an ice-free Antarctica is to place it in a warmer zone of the globe. When you move it up to a warmer climate (so it's mappable) it then becomes the stuff of Atlantis legend. One creative scholar when confronted with these possibilities suggested a comparison of the surface of the earth with the skin of a navel orange, he suggested that a 1/8th slice of the earth simply rotated—carrying Atlantis to its present location as Antarctica, from the middle of the Atlantic Ocean. My homespun logic applied to this bizarre vision, asks the relevant question—where's the force driving this exchange?

The answer is gravity. Specifically the Sun's gravitational pull that makes the Earth bulge by eleven additional miles at the equator and shrink by the same amount at the poles. The only candidate for a compensatory swap that I can see is Australia *(and here I will refrain from boxing your ears with my bizarre suggestion that North and South America were further north in the recent glacial past.)* and the theorist doesn't follow his own logic to search for signs of recent ice movement on Australian soil.

Weight Displacement Potential: Slightly higher during the Galactic Plane Crossing around 2012

If there was such evidence to be found—then it would be time to raise this spectral terror—If Antarctica located on the South Pole was dislocated by the weight of an ice burden toward the equator? What then is the critical height of ice needed to impel or make possible the *pole to equator swap*? The common-sense answer is probably 5 ½ miles—but what if it is only 2 3/4 miles? I only bring this up to add substantial scientific interest to the question of the ultimate destination of the inevitable increase in Antarctic storms. *This by the way is clearly my hype. If the 2012, Dec 21 Galactic Plane crossing coincides with the alignment of the planets on the correct (probably far) side of the sun and our earth is isolated alone on its nearest approach to the sun and if the moon is also on the sunny side (new moon) could that alignment produce an exceptional amount of attraction thus bringing the earth inside its normally variable distance of 91,000,000 to 94,500,000 million miles from the sun ? If that unusual level of attraction were paired with the Black Hole and its massive though distant gravitic force also on the farside of the sun, could it be that a storm-heightened Antarctic Icecap would threaten to destabilize in a substantial way?* {In order to defuse the incipient hysteria likely to attend this train of climate thought:} **The pictures of past climates–as incomplete as they are—show almost no ocean levels higher than the one we have now–almost without number are the sketches of past climates with ocean levels lower–from two to six hundred feet–most of those show no significant or recent continental movement as a derivative of increased continental height. Let's lower the level of hype even further. From the internet** *http://www.greatdreams. com/552000.htm* **"According to the scientist astronomer Dr. Harvey Augensen of Northwestern university, he calculated the positions of the nine planets relative to the sun for May 5, 2,000, A very important date . . . At that time an alignment will occur that will place planet Earth all by itself on one side of the sun. Heading directly away from the earth in a straight line on the other side of the sun will be Mercury, mars, Jupiter and Saturn, even the moon and Pluto have a place in this alignment . . . greater disturbances should occur than in 1982." Does ANYBODY recall anything spectacular happening on or about May 5, 2,000? Or 1982? Nevertheless I would not be true to myself if I didn't add the question, Where was Venus at this time? The more pertinent question that has not been mentioned recently is this: How close will the earth approach to the sun on Dec 21, 2012, and will this be its all time nearest approach?** *(This by the way is a gambit–I have neither the depth of study nor the longevity in climate studies to ask this question. But it seems somewhat silly to me to be exposed to extensive journalistic fears of rising oceans, without being aware of a potential trigger for a gigantic Antarctic SPLASH! (By the way, my fears of a gigantic Antarctic splash may have just receded by 2,000,000 miles. My ancient {1940's} understanding of the nearest point of solar approach was 89,000,000 miles. Google via www.GlobeUniversity.edu provided a distance update. "the minimum distance from the sun to* *the earth is (91,000,000 miles), and the maximum distance from the sun to the earth is 94.5 million miles." The same document proclaims. "The gravitational attraction that the Moon exerts on Earth is the major cause of tides in the sea; the Sun has a lesser tidal influence." Elsewhere "the orbit of the Moon is distinctly elliptical with an average eccentricity of 0.0549.) With these facts at my shoulder and my long term commitment to cleanse the ocean by draining oily tanker residue into the four or five major depressions on the earth, The fears of a gigantic Antarctic splash should be reawakened only if the subsiding of the Oceans exceeds by far my prescribed 8" reduction. I stand ready and eager to receive factual reproof on this gambit.)* I was forced into this side of the coin by my so far unsubtantiated (or accepted) theory of dust. In brief, the Dust Bowl, WWII,—and atom bomb testing produced a pattern of ocean fallout that in some of its parts constituted a viable micro-nutrient which spurred a burst of surface murkiness and its consequent fallout as increased precipitation on land. This pattern of increased precipitation beginning in the forties contributed to burying the famous flight of five WWII airplanes at the 262 foot level of the Greenland Ice Cap. Going hand in hand with my soon to be famous theory of the causes of the absolute decline in $CO2$ in the atmosphere for the approximate period of '62 to '64 or '64 to '66. The consequences of this theory should impel the ocean Sherlock's into a determined search to supplement the seeming panacea of iron filings as a sure cure for an overheated globe. The emergence of the volcanic island of Surtsey southwest of Iceland in 1963 may be regarded by some as proof of some kind of consequence related to weight change. The theory in its simplest guise suggests that when Greenland was pressed down—Surtsey rose up. When these weighty theories are stacked up–another weighs the variable influence of atmospheric highs and lows as they congregate in selected areas of the globe, and a much hyped theory of planetary alignment seems to offer some predictive value to future catastrophic events related to weight change— they only add weight to the electrical theories which I attribute to Robert W.

Felix. To wit: The projected tidal weight increase acting on the shores of the Pacific Rim can increasingly provide the necessary catalytic influence for disastrous climate change–particularly as Greenland or Antarctic melting might enlarge the size and scope of the Pacific and hence its tides. *At the very least this argument adds emotional depth to my plea for increased ocean access to the depths of the five major dry depressed zones of the globe. And it might relieve the anxiety of the Netherlands about the forthcoming setback to the newest global climate conference.*

One additional factoid from Univ. Of Bristol (2008, Jan 14), "Increasing Amounts of Ice Mass Have Been Lost from "Overall in the Antarctic Peninsula, seven ice shelves have between them declined in area by about 13,500sq. Km. since 1974." I bring this up to doubly underscore the additional solar horsepower being absorbed around the Antarctic Ice Cap.–Sept. Thru March The contrast between 85% reflective and 80 or 90% absorptive could easily produce hurricane level horsepower–but I won't bore you with the drawn out figures. (What is obvious from these scare stories is that its relatively easy to sit at a desk and study a satellite picture to note decreasing areas of ice coverage–what is not so obvious is any increase in storm deposition which may in the near future trigger a dangerous increase in the height of the Antarctic Ice Cap.)

The obvious paradox of human time coinciding with an ice free Antarctic is one I choose to duck in the same way that an ice—covered North America seems to suffer large mammal extinctions when the mammoths are apparently snuffed in Eurasia. There are many varied time estimates which are in apparent conflict with one another.

The micro—nutrient contribution of volcanoes is formalized in measurements of Pinatubo but its link to increased Pacific phytoplankton can only be guessed at.

Recent news items associated with the Iceland volcano links, phytoplankton increase and CO2 reduction to Pinatubo eruption and other volcanic activity, The deepest ice core–greatest time coverage–(Vostok 3330 m) The presence of volcanic ash at 3311 M has led to the restriction on the discussion of deeper material. But they discussed it anyway " A sudden decrease from interglacial-like (ie. altered deuterium) ice values to glacial-like values, followed by an abrupt return to interglacial values, occurs between 3320 and 3330 m {Petit et al. 1999} This occurrence plus the presence of volcanic ash layers at 3311m suggests that the Vostok climate records may be disturbed below 3311m. It is worth underlining that Worldwide precipitation patterns can be enormously altered by volcanism either above or below sea level. While we are on this subject it is worth noting that Mt Pinatubo (2nd largest volcano of the 20th century–1st is Novarupto Nov 1 1912 on the Alaskan Peninsula–) Pinatubo is catalogued for elemental contribution to the ocean, with 2.5 cu miles of emissions also emitted 800,000 tons of zinc, 600,000 tons of copper, 600,000 tons of chromium, 300,000 tons of nickel and 100,000 tons of lead among other lesser amounts. These figures are taken from page 109 of 'NATURE'S FURY' LIFE 08 by Time Inc. These estimates are a vitally needed effort to aid the 'Ocean Sherlocks' (CLIMATOLOGISTS) in their search for the vitally necessary micro-nutrient to redeem ocean health. Do I need to re-emphasize that the overwhelming Mid-American floods of '93 can be partly traced to the eruption of Mt Pinatubo? (1991)

Exxon-Mobil as unlikely savior of oceanic phytoplankton?

And to stretch this bit of data into a small sermonette, I need to add that the Japanese have a saying about volcanoes, "Bringer of Life" The vignette is completed when a nameless observer of the oceans chimes in with the observation. Pacific Phytoplankton has been increasing. We have already noted that The Perfect Storm calls attention to an Atlantic Decrease in Phytoplankton. The red flags signaling the shortage of useful knowledge of the state of phytoplankton in the two oceans are highlighted by the recent announcement by Exxon-Mobil that they have succeeded in producing enormously productive algae (for biodiesel) in Colorado, a state that is almost 1,000 miles from Pacific phytoplankton. And we have already noted the 400 hyper-enriched 'Dead zones' led by our own Gulf of Mexico.

There are plenty of unanswered questions raised in this short treatise on water, but none should raise the scientist's lust for knowledge higher than the question of ocean health as measured by phytoplankton. But I believe that I have raised more than one beacon of hope, that can re-energize the world's peoples without the onerous presence of the trappings of Empire. May God bless the readers and actors on this information. Amen.

"For I know that the LORD is great, and our Lord is above all gods. Whatever the LORD pleases He does, in heaven and in earth, in the seas and in all deep places. He causes the vapors to ascend from the ends of the earth; He makes lightning for the rain; He brings the wind out of His treasuries. . ." (Psalm 135: 5-7 NKJV)

INDEX